QUANTUM CHEMISTRY

Edited by
Levitah Castil Mapatac

Delve Publishing

Quantum Chemistry

Edited by Levitah Castil Mapatac

Edition 2018

Delve Publishing

Published by
Delve Publishing LLC
708 3rd Avenue, 6th Floor
New York
NY 10801
United States of America
www.delvepublishing.com

ISBN: 978-1-68095-729-7 (Hardback)

Printed in United States of America

Library of Congress Control Number: 2016920409

About the Editor

Levitah Castil Mapatac

Levitah C. Mapatac is an Associate Professor, researcher, writer and a PhD holder from the University in Science and Technology of Southern Philippines. Her interest varies from different chemistry concepts and applications and currently doing research in traditional medicines, water waste pollution and environmental chemistry in the aspect of contamination pathways, profilling and rehabilatation of mining waste.

Table of Contents

List of Abbreviations xiii
Preface xv

1 PARTICLE IN A BOX

Introduction 1
One-Dimensional Solution........ 3
Position Wave Function........ 3
Momentum Wave Function 6
Position and Momentum Probability Distributions........ 6
Energy Levels........ 8
Higher-Dimensional Boxes 9
Applications 10
Quantum Well Laser 11
Quantum Dots 12
References........ 13

2 WAVE–PARTICLE DUALITY

Introducton........ 17
Brief History of Wave and Particle Viewpoints 18
Turn of the 20Th Century and the Paradigm Shift 21
Particles of electricity 21
Radiation Quantization 21
Photoelectric Effect Illuminated........ 22
Einstein›s explanation of the Photoelectric Effect 23
De Broglie›s Wavelength........ 24
Heisenberg›s Uncertainty Principle 25
de Broglie–Bohm Theory 26
Wave Behavior of Large Objects 27
Treatment in Modern Quantum Mechanics 28
Visualization 28

Alternative Views 29
Both-particle-and-Wave View 29
Wave-only view 30
Particle-only view 31
Neither-wave-nor-Particle View 31
Relational Approach to Wave–Particle Duality 31
Applications 31
References 31

3 VALENCE BOND THEORY

Introduction 38
History 38
Theory 38
Valence Bond Theory Today 39
Applications of Valence Bond Theory 40
Generalized Valence Bond 40
Theory 40
Calculations 41
Modern Valence Bond Theory 41
Spin-Coupled Theory 41
Valence Bond Programs 42
Resonance (Chemistry) 42
General Characteristics of Resonance 43
Misconception 44
Use of Contributing Structures 44
Contributing Structures in Diagrams 45
Examples 46
Electron-Deficient Molecules 47
Reactive Intermediates 47
Bond Lengths 48
Resonance Energy 49
Resonance Energy of Benzene 49
Resonance in Valence Bond (VB) Theory 49
Comparison with Molecular Orbital (MO) Theory 50
Coefficients 51
Charge Delocalization 51
Rferences 52

4 ATOMIC ORBITALS

Introducton....55
Electron Properties....56
Formal Quantum Mechanical Definition....57
Types of Orbitals....58
History....59
Early Models....59
Bohr Atom....60
Modern Conceptions and Connections to the Heisenberg Uncertainty Principle....61
Orbital Names....62
Hydrogen-Like Orbitals....62
Quantum Numbers....63
Complex Orbitals....63
Real Orbitals....64
Shapes of Orbitals....65
Orbitals Table....68
Qualitative Understanding of Shapes....69
Orbital Energy....70
Electron Placement and the Periodic Table....72
Relativistic Effects....73
Transitions Between Orbitals....74
Atoms in Molecules....75
Applications....76
References....77

5 MOLECULAR ORBITAL THEORY

Introduction....82
Quantitative Applications....82
History....82
Types of Orbitals....83
Overview....84
Molecular Orbital....86
Formation of Molecular Orbitals....87
Qualitative Discussion....87
Linear Combinations of Atomic Orbitals (LCAO)....87
Bonding, Antibonding, and Nonbonding MOs....88
Sigma and pi Labels for MOs....88
σ Symmetry....89
π Symmetry....89
δ Symmetry....89

ϕ Symmetry 89
Gerade and Ungerade Symmetry 90
Molecular Orbital Diagram 91
Basics 91
s-p Mixing 93
Diatomic MO Diagrams 94
Dihelium and Diberyllium 95
Diboron 96
Triatomic Molecules 100
Bonding in Molecular Orbitals 104
HOMO and LUMO 104
Molecular Orbital Examples 105
Homonuclear Diatomics 105
H_2 105
He_2 106
Li_2 106
Noble Gases 106
Heteronuclear Diatomics 106
HF 107
Quantitative Approach 107
References 107

6 RELATIVISTIC QUANTUM CHEMISTRY

Introduction 109
History 110
Qualitative Treatment 111
Periodic Table Deviations 113
Mercury 113
Color of Gold and Caesium 114
Inert Pair Effect 115
Others 115
Relativistic Mechanics 115
Relativistic Kinematics 116
Relativistic Dynamics 117
Rest Mass and Relativistic Mass 118
Angular momentum 121
Torque 122
Classical limit 123
References 123

7 SCHRÖDINGER EQUATION

Introduction 126
Equation 127
Time-dependent equation 127
Time-Independent Equation 129
Derivation 130
Implications 130
Total, kinetic, and Potential Energy 130
Quantization 130
Measurement and Uncertainty 131
Quantum Tunneling 131
Particles as Waves 132
Multiverse 133
Interpretation of the Wave Function 134
Historical Background and Development 134
The Wave Equation for Particles 136
Consistency with Energy Conservation 137
Consistency with the De Broglie Relations 138
Wave and Particle Motion 140
Nonrelativistic Quantum Mechanics 142
Time Independent 142
One-Dimensional Examples 143
Free Particle 144
Constant Potential 144
Harmonic Oscillator 145
Three-dimensional Examples 146
Hydrogen Atom 147
Two-electron Atoms or Ions 148
Time Dependent 149
Properties 149
Linearity 150
Real Energy Eigenstates 150
Space and Time Derivatives 151
Local Conservation of Probability 152
Positive Energy 152
Analytic Continuation to Diffusion 153
Regularity 153
Relativistic Quantum Mechanics 154
Quantum Field Theory 155
First Order Form 155

Refeeences 155

8 ELECTRONIC BAND STRUCTURE

Introduction 160
Why Bands And Band Gaps Occur 160
Basic Concepts 161
Crystalline Symmetry and Wavevectors 162
Asymmetry: Band Structures in non-crystalline Solids 163
Filling of Bands 164
Names of bands near the Fermi level (Conduction band, Valence Band) 165
Theory Of Band Structures In Crystals 165
Nearly free Electron Approximation 166
Tight Binding Model 166
KKR Model 167
Density-Functional Theory 167
Green's Function Methods and the ***ab initio*** GW approximation 168
Mott insulators 169
Others 169
Band Diagrams 169
Nearly Free Electron Model 170
Introduction - a Heuristic Argument 170
Mathematical Formulation 172
Results 174
Justifications 174
Tight Binding 175
Historical Background 176
Mathematical Formulation 176
Translational Symmetry and Normalization 177
The tight Binding Hamiltonian 178
The Tight Binding Matrix Elements 178
Evaluation of the Matrix Elements 179
Connection to Wannier functions 180
Second Quantization 180
Example: One-dimensional s-band 181
Table of Interatomic Matrix Elements 182
Muffin-Tin Approximation 184
Density Functional Theory 185
Derivation and Formalism 186
Relativistic Density Functional Theory (explicit Functional Forms) 188
Approximations (exchange-Correlation Functionals) 189
Generalizations to Include Magnetic Fields 190

Applications 190
Thomas–Fermi model 191
Hohenberg–Kohn theorems 192
Pseudo-Potentials 193
Ab initio Pseudo-potentials 193
Electron Smearing 194
K·P Perturbation Theory 194
Background and Derivation 194
Perturbation Theory 195
Expression for a nondegenerate band 195
Effective Mass 196
k·p model with spin-orbit interaction 196
Empty Lattice Approximation 197
Scattering and Periodicity 197
The Energy Bands and the Density of States 198
Second, Third and Higher Brillouin zones 199
The Nearly Free Electron Model 199
The Electron Bands of Common Metal Crystals 199
Refernces 202

9 MOLECULAR HAMILTONIAN

Intoduction 208
Coulomb Hamiltonian 209
Small Terms 210
The Schrödinger Equation of the Coulomb Hamiltonian 211
Clamped Nucleus Hamiltonian 211
Harmonic Nuclear Motion Hamiltonian 212
Watson's Nuclear Motion Hamiltonian 214
References 216

10 BORN–HUANG APPROXIMATION

Introduction 217
Mathematical Formula 218
Consequences 218
Upper-Bound Property 218
Vibronic Coupling 218
Evaluation 219
Crossings and Avoided Crossings of Potential Energy Surfaces 220
Difficulties and Alternatives 220
Theoretical Development 221

Conical Intersection 221
Local Characterization of Conical Intersections 222
Categorization by Symmetry of Intersecting Electronic States 223
References 224

11 APPLICATIONS OF ROTATIONAL SPECTROSCOPY

Introduction 225
Applications 226
Overview 227
Classification of Molecular Rotors 228
Selection Rules 229
Raman Spectra 229
Units 230
Effect of Vibration on Rotation 230
Effect of Rotation on Vibrational Spectra 230
Structure Of Rotational Spectra 231
Spherical top 231
Linear Molecules 232
Rotational line Intensities 233
Centrifugal Distortion 234
Quadrupole Splitting 236
Stark And Zeeman Effects 237
Rotational Raman Spectroscopy 237
Instruments And Methods 237
Absorption cells and Stark Modulation 238
Fourier Transform Microwave (FTMW) Spectroscopy 238
The Balle-Flygare FTMW Spectrometer 238
Rotational–Vibrational Spectroscopy 239
Linear Molecules 241
Spherical Top Molecules 248
Symmetric Top Molecules 249
Asymmetric Top Molecules 251
Experimental Methods 252
References 252

Index 255

List of Abbreviations

AO	Atomic Orbital
DFT	Density Functional Theory
HMO	Hückel Molecular Orbital
HF	Hydrogen Fluoride
LCAO	Linear Combinations of Atomic Orbitals,
MWI	Many-worlds Interpretation
MO	Molecular Orbital
NBO	Natural Bond Orbitals
PES	Potential energy surface
QTAIM	Quantum Theory of Atoms in Molecules
SCF	Self-consistent Field
VB	Valence Bond

Preface

Quantum chemistry is a branch of chemistry whose primary focus is the application of quantum mechanics in physical models and experiments of chemical systems. It is also called molecular quantum mechanics. In reactions, quantum chemistry studies the ground state of individual atoms and molecules, the excited states, and the transition states that occur during chemical reactions. On the calculations: quantum chemical studies use also semi-empirical and other methods based on quantum mechanical principles, and deal with time dependent problems. Many quantum chemical studies assume the nuclei are at rest (Born–Oppenheimer approximation). Many calculations involve iterative methods that include self-consistent field methods. Major goals of quantum chemistry include increasing the accuracy of the results for small molecular systems, and increasing the size of large molecules that can be processed, which is limited by scaling considerations—the computation time increases as a power of the number of atoms. The foundation of quantum mechanics and quantum chemistry is the wave model, in which the atom is a small, dense, positively charged nucleus surrounded by electrons. The first step in solving a quantum chemical problem is usually solving the Schrödinger equation with the electronic molecular Hamiltonian. This Text is meant to be introductory and points out the underlying mathematical framework in all of physical sciences. It is a must for all areas of chemistry where atomistic level explanations are needed for understanding concepts. It covers origins of quantum theory, Schrödinger wave equation, quantum mechanics of simple systems, electron spin, quantum states of atoms, Born–Huang approximation, and much more. It shows the reader which parts of the text are needed for understanding specific aspects of the subject matter.

Content Coverage

Chapters One and *Two* are intended to present an overview of particle in a box model and wave-particle duality, correspondingly. The particle in a box model provides one of the very few problems in quantum mechanics which can be solved analytically, without approximations.

Chapter Three focuses on how a covalent bond is formed between the two atoms by the overlap of half-filled valence atomic orbitals of each atom containing one unpaired electron.

Chapters Four and *Five* include facts about the atomic orbitals and molecular orbital theory, respectively. Atomic orbitals are the basic building blocks of the atomic orbital model, a modern framework for visualizing the submicroscopic behavior of electrons in matter. Whereas, molecular orbital (MO) theory is a method for determining molecular structure in which electrons are not assigned to individual bonds between atoms, but are treated as moving under the influence of the nuclei in the whole molecule.

Chapter Six focuses on relativistic quantum chemistry. Relativistic effects in chemistry can be considered to be perturbations, or small corrections, to the non-relativistic theory of chemistry, which is developed from the solutions of the Schrödinger equation.

Chapters Seven and *Eight* are envisioned to focus on Schrödinger equations and electronic band theory, separately. Band theory has been successfully used to explain many physical properties of solids, such as electrical resistivity and optical absorption, and forms the foundation of the understanding of all solid-state devices.

Chapter Nine aims to the Hamiltonian operator that represents the energy of the electrons and nuclei in a molecule. This operator and the associated Schrödinger equation play a central role in computational chemistry and physics for computing properties of molecules and aggregates of molecules, such as thermal conductivity, specific heat, electrical conductivity, optical, and magnetic properties, and reactivity.

Chapter Ten is aimed at the Born–Huang approximation that takes into account diagonal nonadiabatic effects in the electronic Hamiltonian better than the Born–Oppenheimer approximation.

Chapter Eleven emphasizes on the applications of Rotational spectroscopy, which is concerned with the measurement of the energies of transitions between quantized rotational states of molecules in the gas phase.

CHAPTER 1

Particle in a Box

Chapter Contents

- **Introduction**
- **One-Dimensional Solution**

Assume the potential U(x) in the time-independent Schrodinger equation to be zero inside a one-dimensional box of length L and infinite outside the box. For a particle inside the box a free particle wavefunction is appropriate, but since the probability of finding the particle outside the box is zero, the wavefunction must go to zero at the walls. This constrains the form of the solution to

$$\Psi(x) = A \sin kx, \quad \Psi(0) = \Psi(L) = 0$$

which requires $k = \frac{n\pi}{L}$ (Compare to string modes) and after normalization, the wavefunction is

$$\Psi_n(x) = \sqrt{\frac{2}{L}} \sin \frac{n\pi}{L} x \quad n = 1, 2, 3, \ldots$$

INTRODUCTION

In quantum mechanics, the **particle in a box** model (also known as the **infinite potential well** or the **infinite square well**) describes a particle free to move in a small space surrounded by impenetrable barriers. The model is mainly used as a hypothetical example to illustrate the differences between classical and quantum systems. In classical systems, for example a ball trapped inside a large box, the particle can move at any speed within the box and it is no more likely to be found at one position than another. However, when the well becomes very narrow (on the scale of a few nanometers), quantum effects become important. The particle may only occupy certain positive energy levels. Likewise, it can never have zero energy, meaning that the particle can never "sit still". Additionally, it is more likely to be found at certain positions than at others, depending on its energy level. The particle may never be detected at certain positions, known as spatial nodes.

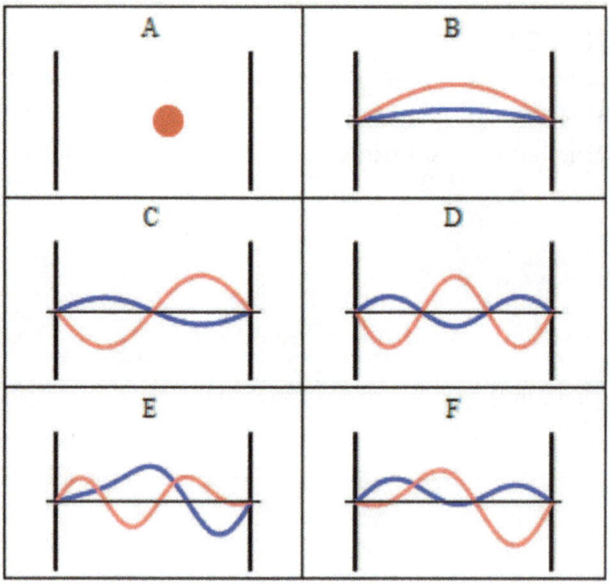

Some trajectories of a particle in a box according to Newton's laws of classical mechanics (A), and according to the Schrödinger equation of quantum mechanics (B-F). In (B-F), the horizontal axis is position, and the vertical axis is the real part (blue) and imaginary part (red) of the wavefunction. The states (B,C,D) are energy eigenstates, but (E,F) are not.

The particle in a box model provides one of the very few problems in quantum mechanics which can be solved analytically, without approximations. This means that the observable properties of the particle (such as its energy and position) are related to the mass of the particle and the width of the well by simple mathematical expressions. Due to its simplicity, the model allows insight into quantum effects without the need for complicated mathematics. It is one of the first quantum mechanics problems taught in undergraduate physics courses, and it is commonly used as an approximation for more complicated quantum systems.

ONE-DIMENSIONAL SOLUTION

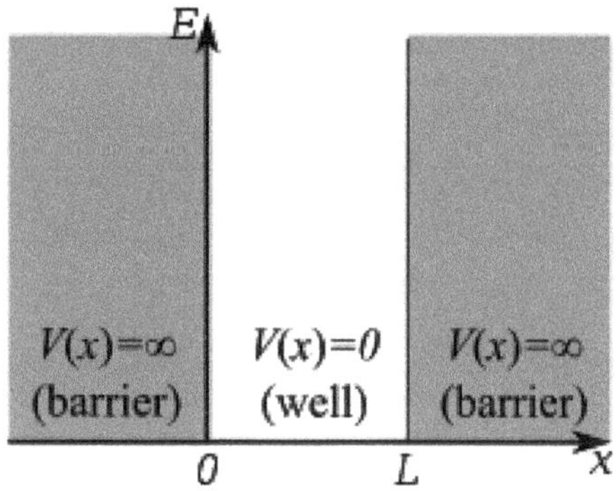

The barriers outside a one-dimensional box have infinitely large potential, while the interior of the box has a constant, zero potential.

The simplest form of the particle in a box model considers a one-dimensional system. Here, the particle may only move backwards and forwards along a straight line with impenetrable barriers at either end. The walls of a one-dimensional box may be visualised as regions of space with an infinitely large potential energy. Conversely, the interior of the box has a constant, zero potential energy. This means that no forces act upon the particle inside the box and it can move freely in that region. However, infinitely large forces repel the particle if it touches the walls of the box, preventing it from escaping. The potential energy in this model is given as

$$V(x) = \begin{cases} 0, & x_c - \frac{L}{2} < x < x_c + \frac{L}{2}, \\ \infty, & \text{otherwise,} \end{cases}$$

where L is the length of the box, x_c is the location of the center of the box and x is the position of the particle within the box. Simple cases include the centered box ($x_c = 0$) and the shifted box ($x_c = L/2$).

Position Wave Function

In quantum mechanics, the wavefunction gives the most fundamental description of the behavior of a particle; the measurable properties of the particle (such as its position, momentum and energy) may all be derived from the wavefunction. The wavefunction $\psi(x,t)$ can be found by solving the Schrödinger equation for the system

$$i\hbar\frac{\partial}{\partial t}\psi(x,t) = -\frac{\hbar^2}{2m}\frac{\partial^2}{\partial x^2}\psi(x,t) + V(x)\psi(x,t),$$

where $\hbar$ is the reduced Planck constant, m is the mass of the particle, i is the imaginary unit and t is time.

Inside the box, no forces act upon the particle, which means that the part of the wavefunction inside the box oscillates through space and time with the same form as a free particle:

$$\psi(x,t) = [A\sin(kx) + B\cos(kx)]\mathrm{e}^{-i\omega t},$$

where A and B are arbitrary complex numbers. The frequency of the oscillations through space and time are given by the wavenumber k and the angular frequency ω respectively. These are both related to the total energy of the particle by the expression

$$E = \hbar\omega = \frac{\hbar^2 k^2}{2m},$$

which is known as the dispersion relation for a free particle. Here one must notice that now, since the particle is not entirely free but under the influence of a potential (the potential V described above), the energy of the particle given above is not the same thing as $\frac{p^2}{2m}$ where p is the momentum of the particle, and thus the wavenumber k above actually describes the energy states of the particle, not the momentum states (id est, it turns out that the momentum of the particle is not given by $p = \hbar k$). In this sense, it is quite dangerous to call the number k a wavenumber, since it is not related to momentum like "wavenumber" usually is. The rationale for calling k the wavenumber is that it enumerates the number of crests that the wavefunction has inside the box, and in this sense it is a wavenumber. This discrepancy can be seen more clearly below, when we find out that the energy spectrum of the particle is discrete (only discrete values of energy are allowed) but the momentum spectrum is continuous (momentum can vary continuously) and in particular, the relation $E = \frac{p^2}{2m}$ for the energy and momentum of the particle does not hold. As said above, the reason this relation between energy and momentum does not hold is that the particle is not free, but there is a potential V in the system, and the energy of the particle is $E = T + V$, where T is the kinetic and V the potential energy.

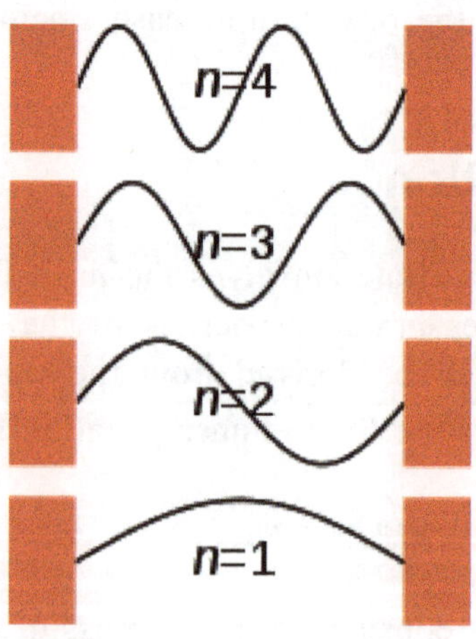

Initial wavefunctions for the first four states in a one-dimensional particle in a box

The size (or amplitude) of the wavefunction at a given position is related to the probability of finding a particle there by $P(x,t) = |\psi(x,t)|^2$. The wavefunction must therefore vanish everywhere beyond the edges of the box. Also, the amplitude of the wavefunction may not "jump" abruptly from one point to the next. These two conditions are only satisfied by wavefunctions with the form

$$\psi_n(x,t) = \begin{cases} A\sin(k_n(x - x_c + \frac{L}{2}))\mathrm{e}^{-i\omega_n t}, & x_c - \frac{L}{2} < x < x_c + \frac{L}{2}, \\ 0, & \text{otherwise,} \end{cases}$$

where

$$k_n = \frac{n\pi}{L},$$

and

$$\omega_n = \frac{n^2\pi^2\hbar}{2mL^2},$$

where n is a positive integer (1,2,3,4...). For a shifted box ($x_c = L/2$), the solution is particularly simple. The simplest solutions, $k_n = 0$ or $A = 0$ both yield the trivial wavefunction $\psi(x) = 0$ both yield the trivial wavefunction $\psi(x) = 0$, which describes a particle that does not exist anywhere in the system. Negative values of n are neglected, since they give wavefunctions identical to the positive n solutions except for a physically unimportant sign change. Here one sees that only a discrete set of energy values and wavenumbers k are allowed for the particle. Usually in quantum mechanics it is also demanded that the derivative of the wavefunction in addition to the wavefunction itself be continuous; here this demand would lead to the only solution being the constant zero function, which is not what we desire, so we give up this demand (as this system with infinite potential can be regarded as a nonphysical abstract limiting case, we can treat it as such and "bend the rules"). Note that giving up this demand means that the wavefunction is not a differentiable function at the boundary of the box, and thus it can be said that the wavefunction does not solve the Schrödinger equation at the boundary points $x = 0$ and $x = L$ (but does solve everywhere else).

Finally, the unknown constant A may be found by normalizing the wavefunction so that the total probability density of finding the particle in the system is 1. It follows that

$$|A| = \sqrt{\frac{2}{L}}.$$

Thus, A may be any complex number with absolute value √(2/L); these different values of A yield the same physical state, so A = √(2/L) can be selected to simplify.

It is expected that the *eigenvalues*, i.e., the energy E_n of the box should be the same regardless of its position in space, but $\psi_n(x,t)$ changes. Notice that $x_c - \frac{L}{2}L$ represents a phase shift in the wave function, This phase shift has no effect when solving the Schrödinger equation, and therefore does not affect the *eigenvalue*.

Momentum Wave Function

The momentum wavefunction is proportional to the Fourier transform of the position wavefunction. With $k = p/\hbar$ (note that this *k*, describing the momentum states, is not the same thing as the k_n above, which described the energy states) the momentum wave function is given by:

$$\phi_n(p,t) = \frac{1}{\sqrt{2\pi\hbar}} \int_{-\infty}^{\infty} \psi_n(x,t) e^{-ikx}\, dx = \sqrt{\frac{L}{\pi\hbar}} \left(\frac{n\pi}{n\pi + kL} \right) \operatorname{sinc}\left(\tfrac{1}{2}(n\pi - kL) \right) e^{-ikx_c}$$

where sinc(.) is the sinc function (sinc(x)=sin(x)/x). for the centered box ($x_c = 0$), the solution is particularly simple since the phase factor on the right is unity. Note that in the problematic case of $n\pi + kL = 0$, the sinc function is also zero, and the limiting case is taken.

It can be seen that the momentum spectrum is continuous, and one can conclude that for energy state described by the k_n wavenumber, the momentum can, when measured, attain also other values than $p = \pm\hbar k_n$, although these are the most likely momentum values (the momentum wavefunction attains its greatest values at these momentum values). Hence one also sees that since the energy is $E_n = \frac{\hbar^2 k_n^2}{2m}$ (since we are talking about the n:th energy eigenstate) the relation $E = \frac{p^2}{2m}$ does not necessarily hold; here *p* is the measured momentum (since the energy eigenstate ψ_n is not a momentum eigenstate, and actually not even a superposition of two momentum eigenstates, as one could imagine from equation **(1)** above, it has no well-defined momentum before measurement).

Position and Momentum Probability Distributions

In classical physics, the particle can be detected anywhere in the box with equal probability. In quantum mechanics, however, the probability density for finding a particle at a given position is derived from the wavefunction as $P(x) = |\psi(x)|^2$ For the particle in a box, the probability density for finding the particle at a given position depends upon its state, and is given by

$$P_n(x,t) = \begin{cases} \frac{L}{2} \sin^2(k_n(x - x_c + \frac{L}{2})), & x_c - L/2 < x < x_c + L/2, \\ 0, & \text{otherwise}, \end{cases}$$

Thus, for any value of *n* greater than one, there are regions within the box for which $P(x) = 0$, indicating that *spatial nodes* exist at which the particle cannot be found.

In quantum mechanics, the average, or expectation value of the position of a particle is given by

$$\langle x \rangle = \int_{-\infty}^{\infty} x P_n(x)\, dx.$$

For the steady state particle in a box, it can be shown that the average position is always $\langle x \rangle = x_c$, regardless of the state of the particle. For a superposition of states, the expectation value of the position will change based on the cross term which is proportional to $\cos(\omega t)$.

The variance in the position is a measure of the uncertainty in position of the particle:

$$\mathrm{Var}(x) = \int_{-\infty}^{\infty} (x - \langle x \rangle)^2 P_n(x)\, dx = \frac{L^2}{12}\left(1 - \frac{6}{n^2\pi^2}\right)$$

The probability density for finding a particle with a given momentum is derived from the wavefunction as $P(x) = |\phi(x)|^2$. As with position, the probability density for finding the particle at a given momentum depends upon its state, and is given by

$$P_n(p) = \frac{L}{\pi\hbar}\left(\frac{n\pi}{n\pi + kL}\right)^2 \operatorname{sinc}^2\left(\tfrac{1}{2}(n\pi - kL)\right)$$

where, again, {\displaystyle k=p/\hbar }. The expectation value for the momentum is then calculated to be zero, and the variance in the momentum is calculated to be:

$$\mathrm{Var}(p) = \left(\frac{\hbar n\pi}{L}\right)^2$$

The uncertainties in position and momentum (Δx and Δp) are defined as being equal to the square root of their respective variances, so that:

$$\Delta x \Delta p = \frac{\hbar}{2}\sqrt{\frac{n^2\pi^2}{3} - 2}$$

This product increases with increasing *n*, having a minimum value for *n=1*. The value of this product for *n=1* is about equal to 0.568 $\hbar$ which obeys the Heisenberg uncertainty principle, which states that the product will be greater than or equal to $\hbar/2$.

Another measure of uncertainty in position is the information entropy of the probability distribution H_x:

$$H_x = \int_{-\infty}^{\infty} P_n(x) \log(P_n(x) x_0)\, dx = \log\left(\frac{2L}{e\, x_0}\right)$$

where x_0 is an arbitrary reference length.

Another measure of uncertainty in momentum is the information entropy of the probability distribution H_p:

$$H_p(n) = \int_{-\infty}^{\infty} P_n(p) \log(P_n(p) p_0)\, dp$$

where p_0 is an arbitrary reference momentum. The integral is difficult to express analytically for general *n*, but in the limit as *n* approaches infinity:

$$\lim_{n->\infty} H_p(n) = \log\left(\frac{4\pi\hbar\, e^{2(1-\gamma)}}{L\, p_0}\right)$$

where γ is Euler›s constant. The quantum mechanical entropic uncertainty principle states that for $x_0\, p_0 = \hbar$

$$H_x + H_p(n) \geq \log(e\,\pi) \approx 2.14473... \text{ (nats)}$$

For $x_0 p_0 = \hbar$, the sum of the position and momentum entropies yields:

$$H_x + H_p(\infty) = \log(8\pi e^{1-2\gamma}) \approx 3.06974... \text{ (nats)}$$

which satisfies the quantum entropic uncertainty principle.

Energy Levels

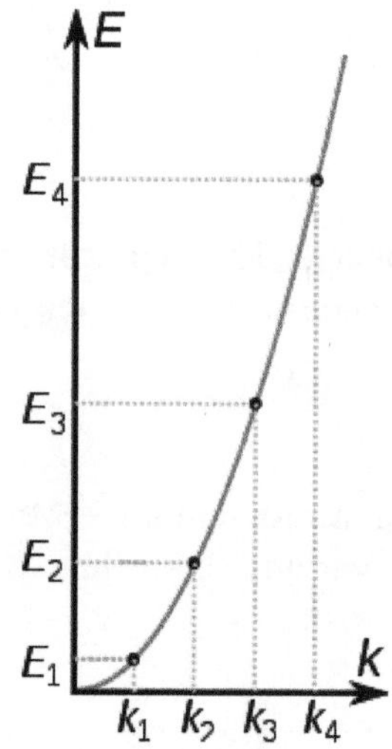

The energy of a particle in a box (black circles) and a free particle (grey line) both depend upon wavenumber in the same way. However, the particle in a box may only have certain, discrete energy levels.

The energies which correspond with each of the permitted wavenumbers may be written as

$$E_n = \frac{n^2\hbar^2\pi^2}{2mL^2} = \frac{n^2h^2}{8mL^2}.$$

The energy levels increase with n^2, meaning that high energy levels are separated from each other by a greater amount than low energy levels are. The lowest possible energy for the particle (its *zero-point energy*) is found in state 1, which is given by

$$E_1 = \frac{\hbar^2\pi^2}{2mL^2}.$$

The particle, therefore, always has a positive energy. This contrasts with classical systems, where the particle can have zero energy by resting motionlessly. This can be explained in terms of the uncertainty principle, which states that the product of the uncertainties in the position and momentum of a particle is limited by

$$\Delta x \Delta p \geq \frac{\hbar}{2}$$

It can be shown that the uncertainty in the position of the particle is proportional to the width of the box. Thus, the uncertainty in momentum is roughly inversely proportional to the width of the box.The kinetic energy of a particle is given by $E = p^2/(2m)$, and hence the minimum kinetic energy of the particle in a box is inversely proportional to the mass and the square of the well width, in qualitative agreement with the calculation above.

Higher-Dimensional Boxes

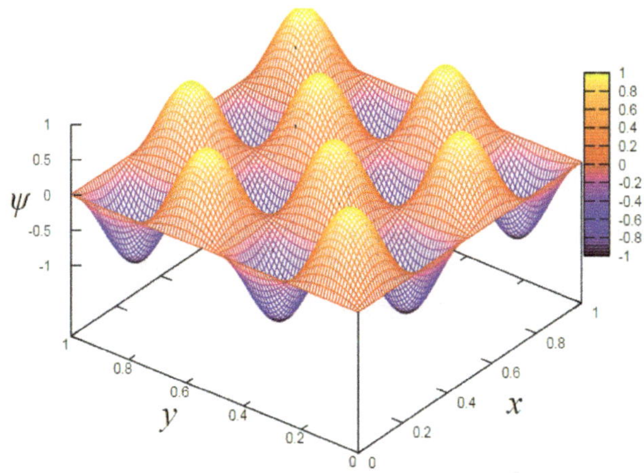

The wavefunction of a 2D well with n_x=4 and n_y=4

If a particle is trapped in a two-dimensional box, it may freely move in the x and y-directions, between barriers separated by lengths L_x and L_y respectively. For a centered box, the position wave function may be written including the length of the box as $\psi_n(x,t,L)$. Using a similar approach to that of the one-dimensional box, it can be shown that the wavefunctions and energies for a centered box are given respectively by

$$\psi_{n_x,n_y} = \psi_{n_x}(x,t,L_x)\psi_{n_y}(y,t,L_y),$$

$$E_{n_x,n_y} = \frac{\hbar^2 k^2_{n_x,n_y}}{2m},$$

where the two-dimensional wavevector is given by

$$\mathbf{k_{n_x,n_y}} = k_{n_x}\hat{\mathbf{x}} + k_{n_y}\hat{\mathbf{y}} = \frac{n_x\pi}{L_x}\hat{\mathbf{x}} + \frac{n_y\pi}{L_y}\hat{\mathbf{y}}.$$

For a three dimensional box, the solutions are

$$\psi_{n_x,n_y,n_z} = \psi_{n_x}(x,t,L_x)\psi_{n_y}(y,t,L_y)\psi_{n_z}(z,t,L_z),$$

$$E_{n_x,n_y,n_z} = \frac{\hbar^2 k^2_{n_x,n_y,n_z}}{2m},$$

where the three-dimensional wavevector is given by:

$$\mathbf{k_{n_x,n_y,n_z}} = k_{n_x}\hat{\mathbf{x}} + k_{n_y}\hat{\mathbf{y}} + k_{n_z}\hat{\mathbf{z}} = \frac{n_x\pi}{L_x}\hat{\mathbf{x}} + \frac{n_y\pi}{L_y}\hat{\mathbf{y}} + \frac{n_z\pi}{L_z}\hat{\mathbf{z}}.$$

In general for an n-dimensional box, the solutions are

$$\psi = \prod_i \psi_{n_i}(x_i, t, L_i)$$

The 1-dimensional momentum wave functions may likewise be represented by $\phi_n(x, t, L_x)$ and the momentum wave function for an n-dimensional centered box is then:

$$\phi = \prod_i \phi_{n_i}(k_i, t, L_i)$$

An interesting feature of the above solutions is that when two or more of the lengths are the same (e.g. $L_x = L_y$), there are multiple wavefunctions corresponding to the same total energy. For example, the wavefunction with $n_x = 2, n_y = 1$ has the same energy as the wavefunction with $n_x = 1, n_y = 2$. This situation is called *degeneracy* and for the case where exactly two degenerate wavefunctions have the same energy that energy level is said to be *doubly degenerate*. Degeneracy results from symmetry in the system. For the above case two of the lengths are equal so the system is symmetric with respect to a 90° rotation.

Applications

Because of its mathematical simplicity, the particle in a box model is used to find approximate solutions for more complex physical systems in which a particle is trapped in a narrow region of low electric potential between two high potential barriers. These quantum well systems are particularly important in optoelectronics, and are used in devices such as the quantum well laser, the quantum well infrared photodetector and the quantum-confined Stark effect modulator. It is also used to model a lattice in the Kronig-Penny model and for a finite metal with the free electron approximation.

Conjugated Polyenes

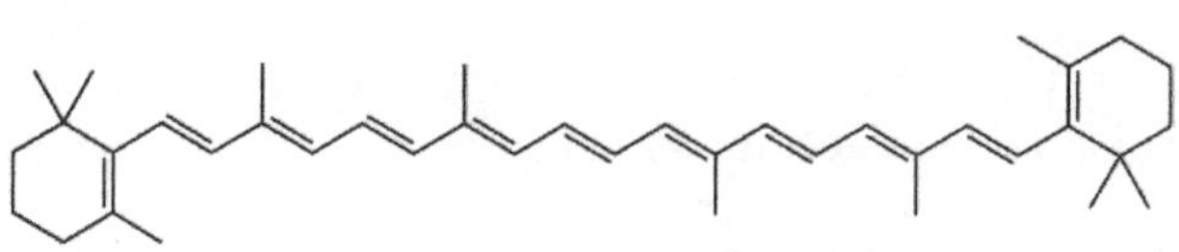

β-carotene is a polyene

Conjugated polyene systems can be modeled using particle in a box. The conjugated system of electrons can be modeled as a one dimensional box with length equal to the total bond distance from one terminus of the polyene to the other. In this case each pair of electrons in each π bond corresponds to one energy level. The energy difference between two energy levels, n_f and n_i is:

$$\Delta E = \frac{(n_f^2 - n_i^2)h^2}{8mL^2}$$

The difference between the ground state energy, n, and the first excited state, n+1, corresponds to the energy required to excite the system. This energy has a specific wavelength, and therefore color of light, related by:

$$\lambda = \frac{hc}{\Delta E}$$

A common example of this phenomenon is in β-carotene β-carotene ($C_{40}H_{56}$) is a conjugated polyene with an orange color and a molecular length of approximately 3.8nm (though its chain length is only approximately 2.4nm). Due to β-carotene's high level of conjugation, electrons are dispersed throughout the length of the molecule, allowing one to model it as a one-dimensional particle in a box. β-carotene has 11 carbon-carbon double bonds in conjugation; each of those double bonds contains two π-electrons, therefore β-carotene has 22 π-electrons. With two electrons per energy level, β-carotene can be treated as a particle in a box at energy level n=11. Therefore, the minimum energy needed to excite an electron to the next energy level can be calculated, n=12, as follows (recalling that the mass of an electron is 9.109×10^{-31} kg):

$$\Delta E = \frac{(n_f^2 - n_i^2)h^2}{8mL^2}$$

$$= \frac{(12^2 - 11^2)h^2}{8mL^2}$$

$$= 2.3658 \times 10^{-19} J$$

Using the previous relation of wavelength to energy, recalling both Planck's constant h and the speed of light c:

$$\lambda = \frac{hc}{\Delta E}$$

$$= 8303au$$

This number describes the wavelength of light, in atomic units, required to induce a transition from beta-carotene's ground state to its first excited state. This can be used to determine the wavelength in nanometers, as 1 atomic unit is 0.529177×10^{-10} m:

$$8303au \times \left(\frac{0.529177 \times 10^{-10} m}{1au}\right) = 4.394 \times 10^{-7} m$$

$$(4.394 \times 10^{-7} m) \times \left(\frac{10^9 nm}{1m}\right) = 439nm$$

The predicted wavelength using this method is 439 nm, while the observed wavelength is 450 nm.

Quantum Well Laser

The particle in a box model can be applied to quantum well lasers, which are laser diodes consisting of one semiconductor "well" material sandwiched between two other semiconductor layers of different material . Because the layers of this sandwich are very thin (the middle layer is typically about 100 Å thick), quantum confinement effects can be observed. The idea that quantum effects could be harnessed to create better laser diodes originated in the 1970s. The quantum well laser was patented in 1976 by R. Dingle and C. H. Henry

Specifically, the quantum well's behavior can be represented by the particle in a finite well model. Two boundary conditions must be selected. The first is that the wave function must be continuous. Often, the second boundary condition is chosen to be the derivative of the wave function must be continuous across the boundary, but in the case of the quantum well the masses are different on either side of the boundary. Instead, the second boundary condition is chosen to conserve particle flux as $(1/m)d\phi/dz$, which is consistent with experiment. The solution to the finite well particle in a box must be solved numerically, resulting in wave functions that are sine functions inside the quantum well and exponentially decaying functions in the barriers. This quantization of the energy levels of the electrons allows a quantum well laser to emit light more efficiently than conventional semiconductor lasers.

Due to their small size, quantum dots do not showcase the bulk properties of the specified semi-conductor but rather show quantised energy states. This effect is known as the quantum confinement and has led to numerous applications of quantum dots such as the quantum well laser.

Researchers at Princeton University have recently built a quantum well laser which is no bigger than a grain of rice. The laser is powered by a single electron which passes through two quantum dots; a double quantum dot. The electron moves from a state of higher energy, to a state of lower energy whilst emitting photons in the microwave region. These photons bounce off mirrors to create a beam of light; the laser.

The quantum well laser is heavily based on the interaction between light and electrons. This relationship is a key component in quantum mechanical theories which include the De Broglie Wavelength and Particle in a box. The double quantum dot allows scientists to gain full control over the movement of an electron which consequently results in the production of a laser beam.

Quantum Dots

Quantum dots are extremely small semiconductors (on the scale of nanometers). They display quantum confinement in that the electrons cannot escape the "dot", thus allowing particle-in-a-box approximations to be applied. Their behavior can be described by three-dimensional particle-in-a-box energy quantization equations.

The energy gap of a quantum dot is the energy gap between its valence and conduction bands. This energy gap (E(r)) is equal to the band gap of the bulk material (E_{gap}) plus the energy equation derived from particle-in-a-box, which gives the energy for electrons and holes. This can be seen in the following equation, where m^*_e and m^*_h are the effective masses of the electron and hole, r is radius of the dot, and h is Planck›s constant:

$$\triangle E(r) = E_{gap} + \left(\frac{h^2}{8r^2}\right)\left(\frac{1}{m_e^*} + \frac{1}{m_h^*}\right)$$

Hence, the energy gap of the quantum dot is inversely proportional to the square of the "length of the box," i.e. the radius of the quantum dot.

Manipulation of the band gap allows for the absorption and emission of specific wavelengths of light, as energy is inversely proportional to wavelength. The smaller the quantum dot, the larger the band gap and thus the shorter the wavelength absorbed.

Different semiconducting materials are used to synthesize quantum dots of different sizes and therefore emit different wavelengths of light. Materials that normally emit light in the visible region are often used and their sizes are fine-tuned so that certain colors are emitted. Typical substances used to synthesize quantum dots are cadmium (Cd) and selenium (Se). For example, when the electrons of two nanometer CdSe quantum dots relax after excitation, blue light is emitted. Similarly, red light is emitted in four nanometer CdSe quantum dots.

Quantum dots have a variety of functions including but not limited to fluorescent dyes, transistors, LEDs, solar cells, and medical imaging via optical probes.

One function of quantum dots is their use in lymph node mapping, which is feasible due to their unique ability to emit light in the near infrared (NIR) region. Lymph node mapping allows surgeons to track if and where cancerous cells exist.

Quantum dots are useful for these functions due to their emission of brighter light, excitation by a wide variety of wavelengths, and higher resistance to light than other substances.

REFERENCES

1. Davies, p.4
2. Actually, any constant, finite potential V_0 can be specified within the box. This merely shifts the energies of the states by V_0.
3. Davies, p. 1
4. Bransden and Joachain, p. 157
5. Davies p. 5
6. Bransden and Joachain, p.158
7. Majernik, Vladimir; Richterek, Lukas (1997-12-01). "Entropic uncertainty relations for the infinite well". J. Phys A. **30** *(4)*. *Bibcode*:1997JPhA...30L..49M. doi:10.1088/0305-4470/30/4/002. Retrieved 11 February 2016.
8. Majernik, Vladimir; Majernikova, Eva (1998-12-01). "The momentum entropy of the infinite potential well". J. Phys A. **32** *(11)*. *Bibcode*:1999JPhA...32.2207M. doi:10.1088/0305-4470/32/11/013. Retrieved 11 February 2016.
9. Bransden and Joachain, p. 159
10. Davies, p. 15

11. Todd Wimpfheimer, A Particle in a Box Laboratory Experiment Using Everyday Compounds, Journal of Laboratory Chemical Education, Vol. 3 No. 2, 2015, pp. 19-21. doi: 10.5923/j.jlce.20150302.01.

12. Todd Wimpfheimer, A Particle in a Box Laboratory Experiment Using Everyday Compounds, Journal of Laboratory Chemical Education, Vol. 3 No. 2, 2015, pp. 19-21. doi: 10.5923/j.jlce.20150302.01.

13. Pubchem. "beta-carotene | C40H56 - PubChem". pubchem.ncbi.nlm.nih.gov. Retrieved 2016-11-10.

14. Sathish, R. K.; Sidharthan, P. V.; Udayanandan, K. M. "Particle in a Box- A Treasure Island for Undergraduates".

15. P.J. Mohr, B.N. Taylor, and D.B. Newell, "The 2014 CODATA Recommended Values of the Fundamental Physical Constants". This database was developed by J. Baker, M. Douma, and S. Kotochigova. Available: . National Institute of Standards and Technology, Gaithersburg, MD 20899.

16. "The NIST Reference on Constants, Units, and Uncertainty". National Institute of Standard and Technology. Retrieved November 28, 2016.

17. β-Carotene http://www.sigmaaldrich.com/catalog/product/aldrich/855553?lang=en®ion=us (accessed Nov 8, 2016).

18. Zory, Peter (1993). Quantum Well Lasers. San Diego: Academic Press Unlimited.

19. U.S. Patent #3,982,207, issued September 21, 1976, Inventors R. Dingle and C. H. Henry ,"Quantum Effects in Heterostructure Lasers", filed March 7, 1975.

20. Miller, David (1995). Burstein, Elias; Weisbuch, Claude, eds. Confined Electrons and Photons: New Physics and Applications. New York: Plenum Press. pp. 675–702.

21. Miessler, G. L. (2013). Inorganic chemistry (5 ed.). Boston: Pearson. pp. 235–236. ISBN 978-0321811059.

22. Miessler, G. L. (2013). Inorganic chemistry (5 ed.). Boston: Pearson. pp. 235–236. ISBN 978-0321811059.

23. Zandonella, Catherine. "Rice-sized laser, powered one electron at a time, bodes well for quantum computing". Princeton University. Princeton University. Retrieved 8 November 2016.

24. Zandonella, Catherine. "Rice-sized laser, powered one electron at a time, bodes well for quantum computing". Princeton University. Princeton University. Retrieved 8 November 2016.

25. Zandonella, Catherine. "Rice-sized laser, powered one electron at a time, bodes well for quantum computing". Princeton University. Princeton University. Retrieved 8 November 2016.

26. Rice, C.V.; Griffin, G.A. (2008). "Simple Syntheses of CdSe Quantum Dots". Journal of Chemical Education. **85** *(6): 842. Retrieved 5 November 2016.*

27. "Quantum Dots : a True "Particle in a Box" System". PhysicsOpenLab. 20 November 2015. Retrieved 5 November 2016.

28. Overney, René M. "Quantum Confinement" (PDF). University of Washington. Retrieved 5 November 2016.

29. Zahn, Dietrich R.T. "Surface and Interface Properties of Semiconductor Quantum Dots by Raman Spectroscopy" (PDF). Technische Universität Chemnitz. Retrieved 5 November 2016.

30. Bentolila, Laurent A.; Ebenstein, Yuval (2009). "Quantum Dots for In Vivo Small-Animal Imaging". Journal of Nuclear Medicine. **50** *(4): 493–496. Retrieved 5 November 2016.*

31. Alberto, P; Fiolhais, C; Gil, V M S (1996). "Relativistic particle in a box" (PDF). European Journal of Physics. **17**: 19–24. Bibcode:1996EJPh...17...19A. doi:10.1088/0143-0807/17/1/004.

32. Yuchuan Wei(2015). The infinite square well problem in the standard, fractional, and relativistic Quantum Mechanics International Journal of Theoretical and Mathematical Physics 2015; 5(4) pp58-65.

CHAPTER 2

Wave–Particle Duality

Chapter Contents

- **Introducton**
- **Brief History of Wave and Particle Viewpoints**
- **Turn of the 20Th Century and the Paradigm Shift**
- **Wave Behavior of Large Objects**
- **Treatment in Modern Quantum Mechanics**
- **Alternative Views**
- **Applications**

A central concept of quantum mechanics, duality addresses the inadequacy of conventional concepts like "particle" and "wave" to meaningfully describe the behaviour of quantum objects.

The idea of duality is rooted in a debate over the nature of light and matter dating back to the 1600s, when competing theories of light were proposed by Christiaan Huygens and Isaac Newton.

Through the work of Albert Einstein, Louis de Broglie and many others, it is now established that all objects have both wave and particle nature (though this phenomenon is only detectable on small scales, such as with atoms), and that a suitable interpretation of quantum mechanics provides the over-arching theory resolving this ostensible paradox.

INTRODUCTON

Wave–particle duality is the concept that every elementary particle or quantic entity may be partly described in terms not only of particles, but also of waves. It expresses the inability of the classical concepts "particle" or "wave" to fully describe the behavior of quantum-scale objects. As Albert Einstein wrote: "It seems as though we must use sometimes the one theory and sometimes the other, while at times we may use either. We are faced with a new kind of difficulty. We have two contradictory pictures of reality; separately neither of them fully explains the phenomena of light, but together they do".

Through the work of Max Planck, Einstein, Louis de Broglie, Arthur Compton, Niels Bohr and many others, current scientific theory holds that all particles also have a wave nature (and vice versa). This phenomenon has been verified not only for elementary particles, but also for compound particles like atoms and even molecules. For macroscopic particles, because of their extremely short wavelengths, wave properties usually cannot be detected.

Although the use of the wave-particle duality has worked well in physics, the meaning or interpretation has not been satisfactorily resolved; see Interpretations of quantum mechanics.

Niels Bohr regarded the "duality paradox" as a fundamental or metaphysical fact of nature. A given kind of quantum object will exhibit sometimes wave, sometimes particle, character, in respectively different physical settings. He saw such duality as one aspect of the concept of complementarity. Bohr regarded renunciation of the cause-effect relation, or complementarity, of the space-time picture, as essential to the quantum mechanical account.

Werner Heisenberg considered the question further. He saw the duality as present for all quantic entities, but not quite in the usual quantum mechanical account considered by Bohr. He saw it in what is called second quantization, which generates an entirely new concept of fields which exist in ordinary space-time, causality still being visualizable. Classical field values (e.g. the electric and magnetic field strengths of Maxwell) are replaced by an entirely new kind of field value, as considered in quantum field theory. Turning the reasoning around, ordinary quantum mechanics can be deduced as a specialized consequence of quantum field theory.

BRIEF HISTORY OF WAVE AND PARTICLE VIEWPOINTS

Democritus—the original atomist—argued that all things in the universe, including light, are composed of indivisible sub-components (light being some form of solar atom). At the beginning of the 11th Century, the Arabic scientist Alhazen wrote the first comprehensive treatise on optics; describing refraction, reflection, and the operation of a pinhole lens via rays of light traveling from the point of emission to the eye. He asserted that these rays were composed of particles of light. In 1630, René Descartes popularized and accredited the opposing wave description in his treatise on light, showing that the behavior of light could be re-created by modeling wave-like disturbances in a universal medium ("plenum"). Beginning in 1670 and progressing over three decades, Isaac Newton developed and championed his corpuscular hypothesis, arguing that the perfectly straight lines of reflection demonstrated light's particle nature; only particles could travel in such straight lines. He explained refraction by positing that particles of light accelerated laterally upon entering a denser medium. Around the same time, Newton's contemporaries Robert Hooke and

Christiaan Huygens—and later Augustin-Jean Fresnel—mathematically refined the wave viewpoint, showing that if light traveled at different speeds in different media (such as water and air), refraction could be easily explained as the medium-dependent propagation of light waves. The resulting Huygens–Fresnel principle was extremely successful at reproducing light's behavior and was subsequently supported by Thomas Young›s 1803 discovery of double-slit interference. The wave view did not immediately displace the ray and particle view, but began to dominate scientific thinking about light in the mid 19th century, since it could explain polarization phenomena that the alternatives could not.

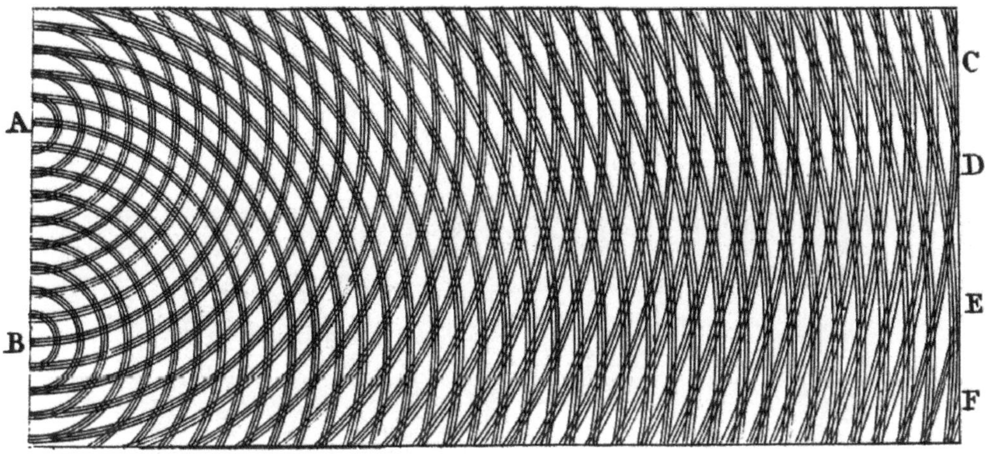

Thomas Young's sketch of two-slit diffraction of waves, 1803.

James Clerk Maxwell discovered that he could apply his equations for electromagnetism, which had been previously discovered, along with a slight modification to describe self-propagating waves of oscillating electric and magnetic fields. When the propagation speed of these electromagnetic waves was calculated, the speed of light fell out. It quickly became apparent that visible light, ultraviolet light, and infrared light (phenomena thought previously to be unrelated) were all electromagnetic waves of differing frequency. The wave theory had prevailed—or at least it seemed to.

While the 19th century had seen the success of the wave theory at describing light, it had also witnessed the rise of the atomic theory at describing matter. Antoine Lavoisier deduced the law of conservation of mass and categorized many new chemical elements and compounds; and Joseph Louis Proust advanced chemistry towards the atom by showing that elements combined in definite proportions. This led John Dalton to propose that elements were invisible sub components; Amedeo Avogadro discovered diatomic gases and completed the basic atomic theory, allowing the correct molecular formulae of most known compounds—as well as the correct weights of atoms—to be deduced and categorized in a consistent manner. Dimitri Mendeleev saw an order in recurring chemical properties, and created a table presenting the elements in unprecedented order and symmetry.

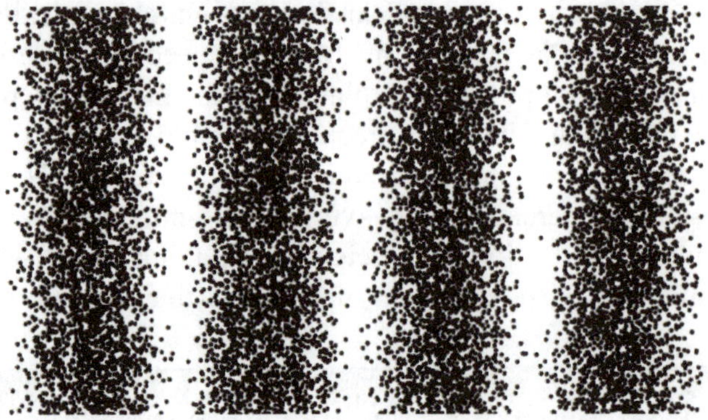

Particle impacts make visible the interference pattern of waves.

A quantum particle is represented by a wave packet.

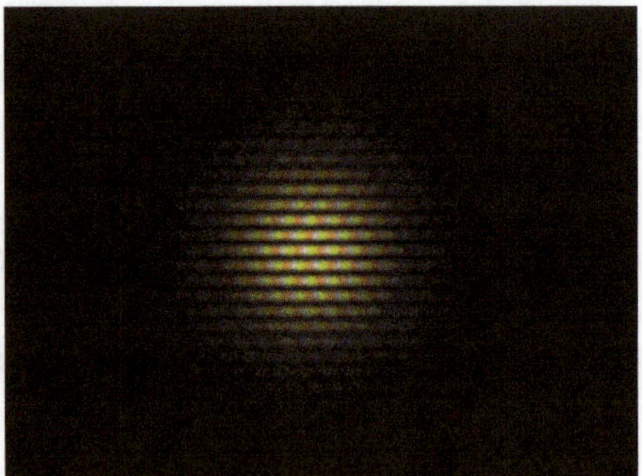

Interference of a quantum particle with itself.

TURN OF THE 20TH CENTURY AND THE PARADIGM SHIFT

Particles of electricity

At the close of the 19th century, the reductionism of atomic theory began to advance into the atom itself; determining, through physics, the nature of the atom and the operation of chemical reactions. Electricity, first thought to be a fluid, was now understood to consist of particles called electrons. This was first demonstrated by J. J. Thomson in 1897 when, using a cathode ray tube, he found that an electrical charge would travel across a vacuum (which would possess infinite resistance in classical theory). Since the vacuum offered no medium for an electric fluid to travel, this discovery could only be explained via a particle carrying a negative charge and moving through the vacuum. This electron flew in the face of classical electrodynamics, which had successfully treated electricity as a fluid for many years (leading to the invention of batteries, electric motors, dynamos, and arc lamps). More importantly, the intimate relation between electric charge and electromagnetism had been well documented following the discoveries of Michael Faraday and James Clerk Maxwell. Since electromagnetism was known to be a wave generated by a changing electric or magnetic field (a continuous, wave-like entity itself) an atomic/particle description of electricity and charge was a non sequitur. Furthermore, classical electrodynamics was not the only classical theory rendered incomplete.

Radiation Quantization

In 1901, Max Planck published an analysis that succeeded in reproducing the observed spectrum of light emitted by a glowing object. To accomplish this, Planck had to make an ad hoc mathematical assumption of quantized energy of the oscillators (atoms of the black body) that emit radiation. Einstein later proposed that electromagnetic radiation itself is quantized, not the energy of radiating atoms.

Black-body radiation, the emission of electromagnetic energy due to an object›s heat, could not be explained from classical arguments alone. The equipartition theorem of classical mechanics, the basis of all classical thermodynamic theories, stated that an object's energy is partitioned equally among the object's vibrational modes. But applying the same reasoning to the electromagnetic emission of such a thermal object was not so successful. That thermal objects emit light had been long known. Since light was known to be waves of electromagnetism, physicists hoped to describe this emission via classical laws. This became known as the black body problem. Since the equipartition theorem worked so well in describing the vibrational modes of the thermal object itself, it was natural to assume that it would perform equally well in describing the radiative emission of such objects. But a problem quickly arose: if each mode received an equal partition of energy, the short wavelength modes would consume all the energy. This became clear when plotting the Rayleigh–Jeans law which, while correctly predicting the intensity of long wavelength emissions, predicted infinite total energy as the intensity diverges to infinity for short wavelengths. This became known as the ultraviolet catastrophe.

In 1900, Max Planck hypothesized that the frequency of light emitted by the black body depended on the frequency of the oscillator that emitted it, and the energy of these oscillators increased linearly with frequency (according to his constant h, where E = hv). This was not an unsound proposal considering that macroscopic oscillators operate similarly: when studying five simple harmonic oscillators of equal amplitude but different frequency, the oscillator with the highest frequency possesses the highest energy (though this relationship is not linear like Planck's). By demanding that high-frequency light must be emitted by an oscillator of equal frequency, and further requiring that this oscillator occupy higher energy than one of a lesser frequency, Planck avoided any catastrophe; giving an equal partition to high-frequency oscillators produced successively fewer oscillators and less emitted light. And as in the Maxwell–Boltzmann distribution, the low-frequency, low-energy oscillators were suppressed by the onslaught of thermal jiggling from higher energy oscillators, which necessarily increased their energy and frequency.

The most revolutionary aspect of Planck›s treatment of the black body is that it inherently relies on an integer number of oscillators in thermal equilibrium with the electromagnetic field. These oscillators give their entire energy to the electromagnetic field, creating a quantum of light, as often as they are excited by the electromagnetic field, absorbing a quantum of light and beginning to oscillate at the corresponding frequency. Planck had intentionally created an atomic theory of the black body, but had unintentionally generated an atomic theory of light, where the black body never generates quanta of light at a given frequency with an energy less than **hv**. However, once realizing that he had quantized the electromagnetic field, he denounced particles of light as a limitation of his approximation, not a property of reality.

Photoelectric Effect Illuminated

While Planck had solved the ultraviolet catastrophe by using atoms and a quantized electromagnetic field, most contemporary physicists agreed that Planck's "light quanta" represented only flaws in his model. A more-complete derivation of black body radiation would yield a fully continuous and 'wave-like' electromagnetic field with no quantization. However, in 1905 Albert Einstein took Planck's black body model to produce his solution to another outstanding problem of the day: the photoelectric effect, wherein electrons are emitted from atoms when they absorb energy from light. Since their discovery eight years previously, electrons had been studied in physics laboratories worldwide.

In 1902 Philipp Lenard discovered that the energy of these ejected electrons did not depend on the intensity of the incoming light, but instead on its frequency. So if one shines a little low-frequency light upon a metal, a few low energy electrons are ejected. If one now shines a very intense beam of low-frequency light upon the same metal, a whole slew of electrons are ejected; however they possess the same low energy, there are merely more of them. The more light there is, the more electrons are ejected. Whereas in order to get high energy electrons, one must illuminate the metal with high-frequency light. Like blackbody radiation, this was at odds with a theory invoking continuous transfer of energy between radiation and matter. However, it can still be explained using a fully classical description of light, as long as matter is quantum mechanical in nature.

If one used Planck's energy quanta, and demanded that electromagnetic radiation at a given frequency could only transfer energy to matter in integer multiples of an energy quantum **hν**, then the photoelectric effect could be explained very simply. Low-frequency light only ejects low-energy electrons because each electron is excited by the absorption of a single photon. Increasing the intensity of the low-frequency light (increasing the number of photons) only increases the number of excited electrons, not their energy, because the energy of each photon remains low. Only by increasing the frequency of the light, and thus increasing the energy of the photons, can one eject electrons with higher energy. Thus, using Planck's constant h to determine the energy of the photons based upon their frequency, the energy of ejected electrons should also increase linearly with frequency; the gradient of the line being Planck's constant. These results were not confirmed until 1915, when Robert Andrews Millikan, who had previously determined the charge of the electron, produced experimental results in perfect accord with Einstein›s predictions. While the energy of ejected electrons reflected Planck›s constant, the existence of photons was not explicitly proven until the discovery of the photon antibunching effect, of which a modern experiment can be performed in undergraduate-level labs. This phenomenon could only be explained via photons, and not through any semi-classical theory (which could alternatively explain the photoelectric effect). When Einstein received his Nobel Prize in 1921, it was not for his more difficult and mathematically laborious special and general relativity, but for the simple, yet totally revolutionary, suggestion of quantized light. Einstein›s "light quanta" would not be called photons until 1925, but even in 1905 they represented the quintessential example of wave-particle duality. Electromagnetic radiation propagates following linear wave equations, but can only be emitted or absorbed as discrete elements, thus acting as a wave and a particle simultaneously.

Einstein›s explanation of the Photoelectric Effect

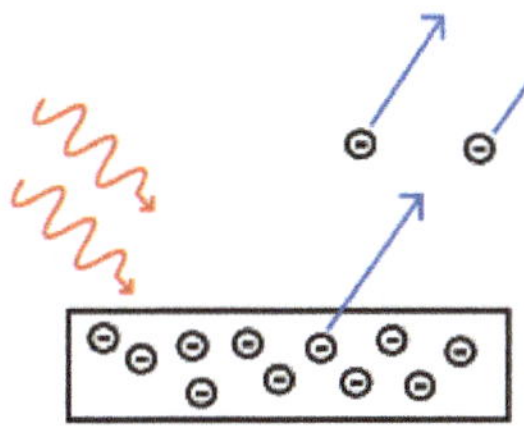

The photoelectric effect. Incoming photons on the left strike a metal plate (bottom), and eject electrons, depicted as flying off to the right.

In 1905, Albert Einstein provided an explanation of the photoelectric effect, a hitherto troubling experiment that the wave theory of light seemed incapable of explaining. He did so by postulating the existence of photons, quanta of light energy with particulate qualities.

In the photoelectric effect, it was observed that shining a light on certain metals would lead to an electric current in a circuit. Presumably, the light was knocking electrons out of the metal, causing current to flow. However, using the case of potassium as an example, it was also observed that while a dim blue light was enough to cause a current, even the strongest, brightest red light available with the technology of the time caused no current at all. According to the classical theory of light and matter, the strength or amplitude of a light wave was in proportion to its brightness: a bright light should have been easily strong enough to create a large current. Yet, oddly, this was not so.

Einstein explained this conundrum by postulating that the electrons can receive energy from electromagnetic field only in discrete portions (quanta that were called photons): an amount of energy E that was related to the frequency f of the light by

$$E = hf$$

where h is Planck›s constant (6.626 × 10^{-34} J seconds). Only photons of a high enough frequency (above a certain threshold value) could knock an electron free. For example, photons of blue light had sufficient energy to free an electron from the metal, but photons of red light did not. One photon of light above the threshold frequency could release only one electron; the higher the frequency of a photon, the higher the kinetic energy of the emitted electron, but no amount of light (using technology available at the time) below the threshold frequency could release an electron. To "violate" this law would require extremely high-intensity lasers which had not yet been invented. Intensity-dependent phenomena have now been studied in detail with such lasers.

Einstein was awarded the Nobel Prize in Physics in 1921 for his discovery of the law of the photoelectric effect.

De Broglie›s Wavelength

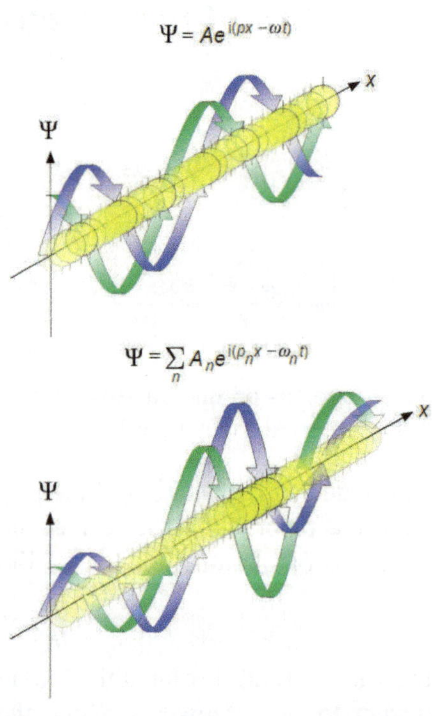

Propagation of de Broglie waves in 1d—real part of the complex amplitude is blue, imaginary part is green. The probability (shown as the colour opacity) of finding the particle at a given point x is spread out like a waveform; there is no definite position of the particle. As the amplitude increases above zero the curvature decreases, so the amplitude decreases again, and vice versa—the result is an alternating amplitude: a wave. Top: Plane wave. Bottom: Wave packet.

In 1924, Louis-Victor de Broglie formulated the de Broglie hypothesis, claiming that all matter, not just light, has a wave-like nature; he related wavelength (denoted as λ), and momentum (denoted as p):

$$\lambda = \frac{h}{p}$$

This is a generalization of Einstein's equation above, since the momentum of a photon is given by $p = \frac{E}{c}$ and the wavelength (in a vacuum) by λ = $\lambda = \frac{c}{f}$, where c is the speed of light in vacuum.

De Broglie›s formula was confirmed three years later for electrons (which differ from photons in having a rest mass) with the observation of electron diffraction in two independent experiments. At the University of Aberdeen, George Paget Thomson passed a beam of electrons through a thin metal film and observed the predicted interference patterns. At Bell Labs, Clinton Joseph Davisson and Lester Halbert Germerguided their beam through a crystalline grid.

De Broglie was awarded the Nobel Prize for Physics in 1929 for his hypothesis. Thomson and Davisson shared the Nobel Prize for Physics in 1937 for their experimental work.

Heisenberg›s Uncertainty Principle

In his work on formulating quantum mechanics, Werner Heisenberg postulated his uncertainty principle, which states:

$$\Delta x \Delta p \geq \frac{\hbar}{2}$$

where

Δ here indicates standard deviation, a measure of spread or uncertainty;

x and **p** are a particle's position and linear momentum respectively.

$\hbar$ is the reduced Planck›s constant (Planck's constant divided by 2π).

Heisenberg originally explained this as a consequence of the process of measuring: Measuring position accurately would disturb momentum and vice versa, offering an example (the "gamma-ray microscope") that depended crucially on the de Broglie hypothesis. The thought is now, however, that this only partly explains the phenomenon, but that the uncertainty also exists in the particle itself, even before the measurement is made.

In fact, the modern explanation of the uncertainty principle, extending the Copenhagen interpretation first put forward by Bohr and Heisenberg, depends even more centrally on the wave nature of a particle: Just as it is nonsensical to discuss the precise location of a wave on a string, particles do not have perfectly precise positions; likewise, just as it is nonsensical to discuss the wavelength of a "pulse" wave traveling down a string, particles do not have perfectly precise momenta (which corresponds to the inverse of wavelength). Moreover, when position is relatively well defined, the wave is pulse-like and has a very ill-defined wavelength (and thus momentum). And conversely, when momentum (and thus wavelength) is relatively well defined, the wave looks long and sinusoidal, and therefore it has a very ill-defined position.

de Broglie–Bohm Theory

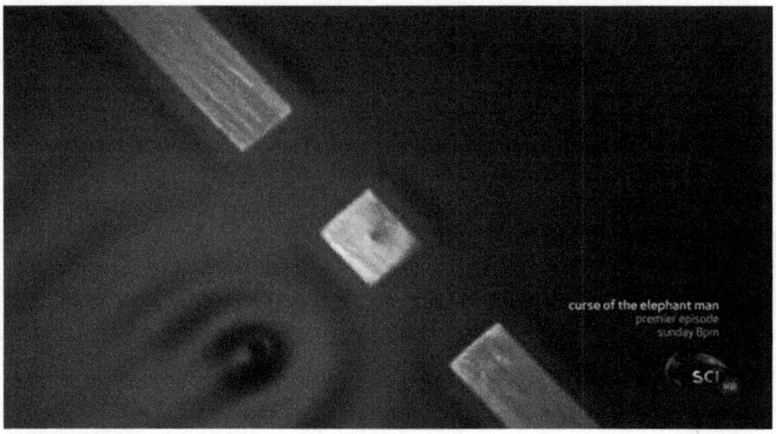

Couder experiments,"materializing" the pilot wave model.

De Broglie himself had proposed a pilot wave construct to explain the observed wave-particle duality. In this view, each particle has a well-defined position and momentum, but is guided by a wave function derived from Schrödinger›s equation. The pilot wave theory was initially rejected because it generated non-local effects when applied to systems involving more than one particle. Non-locality, however, soon became established as an integral feature of quantum theory (see EPR paradox), and David Bohm extended de Broglie's model to explicitly include it.

In the resulting representation, also called the de Broglie–Bohm theory or Bohmian mechanics, the wave-particle duality vanishes, and explains the wave behaviour as a scattering with wave appearance, because the particle›s motion is subject to a guiding equation or quantum potential. "This idea seems to me so natural and simple, to resolve the wave-particle dilemma in such a clear and ordinary way, that it is a great mystery to me that it was so generally ignored", J.S.Bell.

The best illustration of the pilot-wave model was given by Couder›s 2010 "walking droplets" experiments, demonstrating the pilot-wave behaviour in a macroscopic mechanical analog.

WAVE BEHAVIOR OF LARGE OBJECTS

Since the demonstrations of wave-like properties in photons and electrons, similar experiments have been conducted with neutrons and protons. Among the most famous experiments are those of Estermann and Otto Stern in 1929. Authors of similar recent experiments with atoms and molecules, described below, claim that these larger particles also act like waves. A wave is basically a group of particles which moves in a particular form of motion, i.e. to and fro. If we break that flow by an object it will convert into radiants.

A dramatic series of experiments emphasizing the action of gravity in relation to wave–particle duality was conducted in the 1970s using the neutron interferometer. Neutrons, one of the components of the atomic nucleus, provide much of the mass of a nucleus and thus of ordinary matter. In the neutron interferometer, they act as quantum-mechanical waves directly subject to the force of gravity. While the results were not surprising since gravity was known to act on everything, including light (see tests of general relativity and the Pound–Rebka falling photon experiment), the self-interference of the quantum mechanical wave of a massive fermion in a gravitational field had never been experimentally confirmed before.

In 1999, the diffraction of C_{60} fullerenes by researchers from the University of Vienna was reported.Fullerenes are comparatively large and massive objects, having an atomic mass of about 720 u. The de Broglie wavelength of the incident beam was about 2.5 pm, whereas the diameter of the molecule is about 1 nm, about 400 times larger. In 2012, these far-field diffraction experiments could be extended to phthalocyanine molecules and their heavier derivatives, which are composed of 58 and 114 atoms respectively. In these experiments the build-up of such interference patterns could be recorded in real time and with single molecule sensitivity.

In 2003, the Vienna group also demonstrated the wave nature of tetraphenylporphyrin—a flat biodye with an extension of about 2 nm and a mass of 614 u. For this demonstration they employed a near-field Talbot Lau interferometer. In the same interferometer they also found interference fringes for $C_{60}F_{48}$, a fluorinated buckyball with a mass of about 1600 u, composed of 108 atoms. Large molecules are already so complex that they give experimental access to some aspects of the quantum-classical interface, i.e., to certain decoherence mechanisms. In 2011, the interference of molecules as heavy as 6910 u could be demonstrated in a Kapitza–Dirac–Talbot–Lau interferometer. In 2013, the interference of molecules beyond 10,000 u has been demonstrated.

Whether objects heavier than the Planck mass (about the weight of a large bacterium) have a de Broglie wavelength is theoretically unclear and experimentally unreachable; above the Planck mass a particle's Compton wavelength would be smaller than the Planck length and its own Schwarzschild radius, a scale at which current theories of physics may break down or need to be replaced by more general ones.

Recently Couder, Fort, et al. showed that we can use macroscopic oil droplets on a vibrating surface as a model of wave–particle duality—localized droplet creates periodical waves around and interaction with them leads to quantum-like phenomena: interference in double-slit experiment, unpredictable tunneling (depending in complicated way on practically hidden state of field), orbit quantization(that particle has to ‹find a resonance› with field perturbations it creates—after one orbit, its internal phase has to return to the initial state) and Zeeman effect.

TREATMENT IN MODERN QUANTUM MECHANICS

Wave–particle duality is deeply embedded into the foundations of quantum mechanics. In the formalism of the theory, all the information about a particle is encoded in its wave function, a complex-valued function roughly analogous to the amplitude of a wave at each point in space. This function evolves according to a differential equation (generically called the Schrödinger equation). For particles with mass this equation has solutions that follow the form of the wave equation. Propagation of such waves leads to wave-like phenomena such as interference and diffraction. Particles without mass, like photons, have no solutions of the Schrödinger equation so have another wave.

The particle-like behavior is most evident due to phenomena associated with measurement in quantum mechanics. Upon measuring the location of the particle, the particle will be forced into a more localized state as given by the uncertainty principle. When viewed through this formalism, the measurement of the wave function will randomly "collapse", or rather "decohere", to a sharply peaked function at some location. For particles with mass the likelihood of detecting the particle at any particular location is equal to the squared amplitude of the wave function there. The measurement will return a well-defined position, (subject to uncertainty), a property traditionally associated with particles. It is important to note that a measurement is only a particular type of interaction where some data is recorded and the measured quantity is forced into a particular eigenstate. The act of measurement is therefore not fundamentally different from any other interaction.

Following the development of quantum field theory the ambiguity disappeared. The field permits solutions that follow the wave equation, which are referred to as the wave functions. The term particle is used to label the irreducible representations of the Lorentz group that are permitted by the field. An interaction as in a Feynman diagram is accepted as a calculationally convenient approximation where the outgoing legs are known to be simplifications of the propagation and the internal lines are for some order in an expansion of the field interaction. Since the field is non-local and quantized, the phenomena which previously were thought of as paradoxes are explained. Within the limits of the wave-particle duality the quantum field theory gives the same results.

Visualization

There are two ways to visualize the wave-particle behaviour: by the "standard model", described below; and by the Broglie–Bohm model, where no duality is perceived.

Below is an illustration of wave–particle duality as it relates to De Broglie's hypothesis and Heisenberg's uncertainty principle (above), in terms of the position and momentum space wavefunctions for one spinless particle with mass in one dimension. These wavefunctions are Fourier transforms of each other.

The more localized the position-space wavefunction, the more likely the particle is to be found with the position coordinates in that region, and correspondingly the momentum-space wavefunction is less localized so the possible momentum components the particle could have are more widespread.

Conversely the more localized the momentum-space wavefunction, the more likely the particle is to be found with those values of momentum components in that region, and correspondingly the less localized the position-space wavefunction, so the position coordinates the particle could occupy are more widespread.

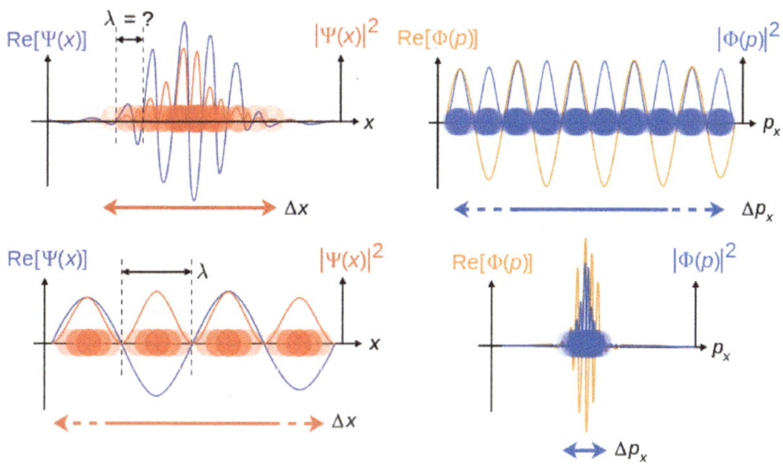

Position x and momentum p wavefunctions corresponding to quantum particles. The colour opacity (%) of the particles corresponds to the probability density of finding the particle with position x or momentum component p.**Top:** If wavelength λ is unknown, so are momentum p, wave-vector k and energy E (de Broglie relations). As the particle is more localized in position space, Δx is smaller than for Δp_x.**Bottom:** If λ is known, so are p, k, and E. As the particle is more localized in momentum space, Δp is smaller than for Δx.

ALTERNATIVE VIEWS

Wave–particle duality is an ongoing conundrum in modern physics. Most physicists accept wave-particle duality as the best explanation for a broad range of observed phenomena; however, it is not without controversy. Alternative views are also presented here. These views are not generally accepted by mainstream physics, but serve as a basis for valuable discussion within the community.

Both-particle-and-Wave View

The pilot wave model, originally developed by Louis de Broglie and further developed by David Bohm into the hidden variable theory proposes that there is no duality, but rather a system exhibits both particle properties and wave properties simultaneously, and particles are guided, in a deterministic fashion, by the pilot wave (or its "quantum potential") which will direct them to areas of constructive interference in preference to areas of destructive interference. This idea is held by a significant minority within the physics community.

At least one physicist considers the "wave-duality" as not being an incomprehensible mystery. L.E. Ballentine, Quantum Mechanics, A Modern Development, p. 4, explains:

When first discovered, particle diffraction was a source of great puzzlement. Are "particles" really "waves?" In the early experiments, the diffraction patterns were detected holistically by means of a photographic plate, which could not detect individual particles. As a result, the notion grew that particle and wave properties were mutually incompatible, or complementary, in the sense that different measurement apparatuses would be required to observe them. That idea, however, was only an unfortunate generalization from a technological limitation. Today it is possible to detect the arrival of individual electrons, and to see the diffraction pattern emerge as a statistical pattern made up of many small spots (Tonomura et al., 1989). Evidently, quantum particles are indeed particles, but whose behaviour is very different from classical physics would have us to expect.

It has been claimed that the Afshar experiment (2007) shows that it is possible to simultaneously observe both wave and particle properties of photons. This claim is, however, rejected by other scientists.

Wave-only view

At least one scientist proposes that the duality can be replaced by a "wave-only" view. In his book Collective Electrodynamics: Quantum Foundations of Electromagnetism (2000), Carver Mead purports to analyze the behavior of electrons and photons purely in terms of electron wave functions, and attributes the apparent particle-like behavior to quantization effects and eigenstates. According to reviewer David Haddon:

Mead has cut the Gordian knot of quantum complementarity. He claims that atoms, with their neutrons, protons, and electrons, are not particles at all but pure waves of matter. Mead cites as the gross evidence of the exclusively wave nature of both light and matter the discovery between 1933 and 1996 of ten examples of pure wave phenomena, including the ubiquitous laser of CD players, the self-propagating electrical currents of superconductors, and the Bose–Einstein condensate of atoms.

Albert Einstein, who, in his search for a Unified Field Theory, did not accept wave-particle duality, wrote:

This double nature of radiation (and of material corpuscles)...has been interpreted by quantum-mechanics in an ingenious and amazingly successful fashion. This interpretation... appears to me as only a temporary way out...

The many-worlds interpretation (MWI) is sometimes presented as a waves-only theory, including by its originator, Hugh Everett who referred to MWI as "the wave interpretation".

The Three Wave Hypothesis of R. Horodecki relates the particle to wave. The hypothesis implies that a massive particle is an intrinsically spatially as well as temporally extended wave phenomenon by a nonlinear law.

Particle-only view

Still in the days of the old quantum theory, a pre-quantum-mechanical version of wave–particle duality was pioneered by William Duane, and developed by others including Alfred Landé. Duane explained diffraction of x-rays by a crystal in terms solely of their particle aspect. The deflection of the trajectory of each diffracted photon was explained as due to quantized momentum transfer from the spatially regular structure of the diffracting crystal.

Neither-wave-nor-Particle View

It has been argued that there are never exact particles or waves, but only some compromise or intermediate between them. For this reason, in 1928 Arthur Eddington coined the name "wavicle" to describe the objects although it is not regularly used today. One consideration is that zero-dimensional mathematical points cannot be observed. Another is that the formal representation of such points, the Dirac delta function is unphysical, because it cannot be normalized. Parallel arguments apply to pure wave states. Roger Penrose states:

"Such ‹position states› are idealized wavefunctions in the opposite sense from the momentum states. Whereas the momentum states are infinitely spread out, the position states are infinitely concentrated. Neither is normalizable [...]."

Relational Approach to Wave–Particle Duality

Relational quantum mechanics is developed which regards the detection event as establishing a relationship between the quantized field and the detector. The inherent ambiguity associated with applying Heisenberg›s uncertainty principle and thus wave–particle duality is subsequently avoided.

APPLICATIONS

Although it is difficult to draw a line separating wave–particle duality from the rest of quantum mechanics, it is nevertheless possible to list some applications of this basic idea.

- Wave–particle duality is exploited in electron microscopy, where the small wavelengths associated with the electron can be used to view objects much smaller than what is visible using visible light.
- Similarly, neutron diffraction uses neutrons with a wavelength of about 0.1 nm, the typical spacing of atoms in a solid, to determine the structure of solids.
- Photos are now able to show this dual nature, perhaps this will lead to new ways of examining and recording this behaviour.

REFERENCES

1. Harrison, David (2002). "Complementarity and the Copenhagen Interpretation of Quantum Mechanics". UPSCALE. Dept. of Physics, U. of Toronto. Retrieved 2008-06-21.

2. Walter Greiner (2001). Quantum Mechanics: An Introduction. Springer. ISBN 3-540-67458-6.
3. R. Eisberg & R. Resnick (1985). Quantum Physics of Atoms, Molecules, Solids, Nuclei, and Particles (2nd ed.). John Wiley & Sons. pp. 59–60. ISBN 047187373X. For both large and small wavelengths, both matter and radiation have both particle and wave aspects.... But the wave aspects of their motion become more difficult to observe as their wavelengths become shorter.... For ordinary macroscopic particles the mass is so large that the momentum is always sufficiently large to make the de Broglie wavelength small enough to be beyond the range of experimental detection, and classical mechanics reigns supreme.
4. Kumar, Manjit (2011). Quantum: Einstein, Bohr, and the Great Debate about the Nature of Reality (Reprint ed.). W. W. Norton & Company. pp. 242, 375–376. ISBN 978-0393339888.
5. Bohr, N. (1927/1928). The quantum postulate and the recent development of atomic theory, NatureSupplement April 14 1928, **121**: 580–590.
6. Camilleri, K. (2009). Heisenberg and the Interpretation of Quantum Mechanics: the Physicist as Philosopher, Cambridge University Press, Cambridge UK, ISBN 978-0-521-88484-6.
7. Preparata, G. (2002). An Introduction to a Realistic Quantum Physics, World Scientific, River Edge NJ, ISBN 978-981-238-176-7.
8. Nathaniel Page Stites, M.A./M.S. "Light I: Particle or Wave?," Visionlearning Vol. PHY-1 (3), 2005. http://www.visionlearning.com/library/module_viewer.php?mid=132
9. Young, Thomas (1804). "Bakerian Lecture: Experiments and calculations relative to physical optics". Philosophical Transactions of the Royal Society. ***94***: 1–16. Bibcode:1804RSPT...94....1Y. doi:10.1098/rstl.1804.0001.
10. Thomas Young: The Double Slit Experiment
11. Buchwald, Jed (1989). The Rise of the Wave Theory of Light: Optical Theory and Experiment in the Early Nineteenth Century. Chicago: University of Chicago Press. ISBN 0-226-07886-8. OCLC 18069573.
12. Lamb, Willis E.; Scully, Marlan O. (1968). "The photoelectric effect without photons" (PDF).
13. «Observing the quantum behavior of light in an undergraduate laboratory". American Journal of Physics. ***72***: 1210. Bibcode:2004AmJPh..72.1210T. doi:10.1119/1.1737397.
14. Zhang, Q (1996). "Intensity dependence of the photoelectric effect induced by a circularly polarized laser beam". Physics Letters A. ***216*** (1-5): 125–128. Bibcode:1996PhLA..216..125Z. doi:10.1016/0375-9601(96)00259-9.
15. Donald H Menzel, "Fundamental formulas of Physics", volume 1, page 153; Gives the de Broglie wavelengths for composite particles such as protons and neutrons.

16. Brian Greene, The Elegant Universe, page 104 "all matter has a wave-like character"

17. See this Science Channel production (Season II, Episode VI "How Does The Universe Work?"), presented by Morgan Freeman, https://www.youtube.com/watch?v=W9yWv5dqSKk

18. Bohmian Mechanics, Stanford Encyclopedia of Philosophy.

19. Bell, J. S., "Speakable and Unspeakable in Quantum Mechanics", Cambridge: Cambridge University Press, 1987.

20. Y. Couder, A. Boudaoud, S. Protière, Julien Moukhtar, E. Fort: Walking droplets: a form of wave-particle duality at macroscopic level? , doi:10.1051/epn/2010101, (PDF)

21. Estermann, I.; Stern O. (1930). "Beugung von Molekularstrahlen". Zeitschrift für Physik. ***61*** (1-2): 95–125. Bibcode:1930ZPhy...61...95E. doi:10.1007/BF01340293.

22. R. Colella, A. W. Overhauser and S. A. Werner, Observation of Gravitationally Induced Quantum Interference, Phys. Rev. Lett. **34**, 1472–1474 (1975).

23. Arndt, Markus; O. Nairz; J. Voss-Andreae, C. Keller, G. van der Zouw, A. Zeilinger (14 October 1999). "Wave–particle duality of C_{60}". Nature. ***401*** (6754): 680–682. Bibcode:1999Natur.401..680A. doi:10.1038/44348. PMID 18494170.

24. Juffmann, Thomas; et al. (25 March 2012). "Real-time single-molecule imaging of quantum interference". Nature Nanotechnology. Retrieved 27 March 2012.

25. Quantumnanovienna. "Single molecules in a quantum interference movie". Retrieved 2012-04-21.

26. Hackermüller, Lucia; Stefan Uttenthaler; Klaus Hornberger; Elisabeth Reiger; Björn Brezger; Anton Zeilinger; Markus Arndt (2003). "The wave nature of biomolecules and fluorofullerenes". Phys. Rev. Lett. ***91*** (9): 090408. arXiv:quant-ph/0309016. Bibcode:2003PhRvL..91i0408H. doi:10.1103/PhysRevLett.91.090408. PMID 14525169.

27. Clauser, John F.; S. Li (1994). "Talbot von Lau interefometry with cold slow potassium atoms.". Phys. Rev. A. ***49*** (4): R2213–17. Bibcode:1994PhRvA..49.2213C. doi:10.1103/PhysRevA.49.R2213. PMID 9910609.

28. Brezger, Björn; Lucia Hackermüller; Stefan Uttenthaler; Julia Petschinka; Markus Arndt; Anton Zeilinger (2002). "Matter-wave interferometer for large molecules". Phys. Rev. Lett. ***88*** (10): 100404. arXiv:quant-ph/0202158. Bibcode:2002PhRvL..88j0404B. doi:10.1103/PhysRevLett.88.100404. PMID 11909334.

29. Hornberger, Klaus; Stefan Uttenthaler; Björn Brezger; Lucia Hackermüller; Markus Arndt; Anton Zeilinger (2003). "Observation of Collisional Decoherence in Interferometry". Phys. Rev. Lett. ***90*** (16): 160401. arXiv:quant-ph/0303093. Bibcode:2003PhRvL..90p0401H. doi:10.1103/PhysRevLett.90.160401. PMID 12731960.

30. Hackermüller, Lucia; Klaus Hornberger; Björn Brezger; Anton Zeilinger; Markus Arndt (2004). "Decoherence of matter waves by thermal emission of radiation". Nature.

427 (6976): 711–714. arXiv:quant-ph/0402146. Bibcode:2004Natur.427..711H. doi:10.1038/nature02276. PMID 14973478.

31. Gerlich, Stefan; et al. (2011). "Quantum interference of large organic molecules". Nature Communications. *2* (263). Bibcode:2011NatCo...2E.263G. doi:10.1038/ncomms1263. PMC 3104521. PMID 21468015.
32. Eibenberger, S.; Gerlich, S.; Arndt, M.; Mayor, M.; Tüxen, J. (2013). "Matter-wave interference of particles selected from a molecular library with masses exceeding 10 000 amu". Physical Chemistry Chemical Physics. *15* (35): 14696–14700. doi:10.1039/c3cp51500a. PMID 23900710.
33. Peter Gabriel Bergmann, The Riddle of Gravitation, Courier Dover Publications, 1993 ISBN 0-486-27378-4online
34. https://www.youtube.com/watch?v=W9yWv5dqSKk - You Tube video - Yves Couder Explains Wave/Particle Duality via Silicon Droplets
35. Y. Couder, E. Fort, Single-Particle Diffraction and Interference at a Macroscopic Scale, PRL 97, 154101 (2006) online
36. A. Eddi, E. Fort, F. Moisy, Y. Couder, Unpredictable Tunneling of a Classical Wave–Particle Association, PRL 102, 240401 (2009)
37. Fort, E.; Eddi, A.; Boudaoud, A.; Moukhtar, J.; Couder, Y. (2010). "Path-memory induced quantization of classical orbits". PNAS. *107* (41): 17515–17520. doi:10.1073/pnas.1007386107.
38. http://prl.aps.org/abstract/PRL/v108/i26/e264503 - Level Splitting at Macroscopic Scale
39. (Buchanan pp. 29–31)
40. Afshar S.S. et al: Paradox in Wave Particle Duality. Found. Phys. 37, 295 (2007) http://arxiv.org/abs/quant-ph/0702188 arXiv:quant-ph/0702188
41. David Haddon. "Recovering Rational Science". Touchstone. Retrieved 2007-09-12.
42. Paul Arthur Schilpp, ed, Albert Einstein: Philosopher-Scientist, Open Court (1949), ISBN 0-87548-133-7, p 51.
43. See section VI(e) of Everett›s thesis: The Theory of the Universal Wave Function, in Bryce Seligman DeWitt, R. Neill Graham, eds, The Many-Worlds Interpretation of Quantum Mechanics, Princeton Series in Physics, Princeton University Press (1973), ISBN 0-691-08131-X, pp 3–140.
44. Horodecki, R. (1981). "De broglie wave and its dual wave". Phys. Lett. A. *87* (3): 95–97. Bibcode:1981PhLA...87...95H. doi:10.1016/0375-9601(81)90571-5.
45. Horodecki, R. (1983). "Superluminal singular dual wave". Lett. Novo Cimento. *38*: 509–511.
46. Duane, W. (1923). The transfer in quanta of radiation momentum to matter, Proc. Natl. Acad. Sci. **9**(5): 158–164.
47. Landé, A. (1951). Quantum Mechanics, Sir Isaac Pitman and Sons, London, pp. 19–22.

48. Heisenberg, W. (1930). The Physical Principles of the Quantum Theory, translated by C. Eckart and F.C. Hoyt, University of Chicago Press, Chicago, pp. 77–78.

49. Eddington, Arthur Stanley (1928). The Nature of the Physical World. Cambridge, UK.: MacMillan. p. 201.

50. Penrose, Roger (2007). The Road to Reality: A Complete Guide to the Laws of the Universe. Vintage. p. 521, §21.10. ISBN 978-0-679-77631-4.

51. http://www.quantum-relativity.org/Quantum-Relativity.pdf. See Q. Zheng and T. Kobayashi, Quantum Optics as a Relativistic Theory of Light; Physics Essays 9 (1996) 447. Annual Report, Department of Physics, School of Science, University of Tokyo (1992) 240.

52. Ecole Polytechnique Federale de Lausanne. "The first ever photograph of light as both a particle and wave". Retrieved 15 July 2016

CHAPTER 3

Valence Bond Theory

Chapter Contents

- **Introduction**
- **History**
- **Theory**
- **Valence Bond Theory Today**
- **Applications of Valence Bond Theory**
- **Generalized Valence Bond**
- **Modern Valence Bond Theory**
- **Resonance (Chemistry)**
- **Resonance Energy**

The valence-bond approach considers the overlap of the atomic orbitals (AO) of the participation atoms to form a chemical bond. Due to the overlapping, electrons are localized in the bond region.

The overlapping AOs can be of different types, for example, a sigma bond may be formed by the overlapping the following AOs.

Chemical bonds formed due to overlap of atomic orbitals					
s-s	*s-p*	*s-d*	*p-p*	*p-d*	*d-d*
H-H Li-H	H-C H-N H-F	H-Pd in palladium hydride	C-C P-P S-S	F-S in SF_6	Fe-Fe

However, the atomic orbitals for bonding may not be "pure" atomic orbitals directly from the solution of the Schrodinger Equation. Often, the bonding atomic orbitals have a character of several possible types of orbitals. The methods to get an AO with the proper character for the bonding is called **hybridization**. The resulting atomic orbitals are called **hybridized atomic orbitals** or simply **hybrid orbitals**.

We shall look at the shapes of some hybrid orbitals first, because these shapes determine the shapes of the molecules.

INTRODUCTION

In chemistry, **valence bond (VB) theory** is one of two basic theories, along with molecular orbital (MO) theory, that were developed to use the methods of quantum mechanics to explain chemical bonding. It focuses on how the atomic orbitals of the dissociated atoms combine to give individual chemical bonds when a molecule is formed. In contrast, molecular orbital theory has orbitals that cover the whole molecule.

HISTORY

In 1916, G. N. Lewis proposed that a chemical bond forms by the interaction of two shared bonding electrons, with the representation of molecules as Lewis structures. In 1927 the Heitler–London theory was formulated which for the first time enabled the calculation of bonding properties of the hydrogen molecule H_2 based on quantum mechanical considerations. Specifically, Walter Heitler determined how to use Schrödinger›s wave equation (1926) to show how two hydrogen atom wavefunctions join together, with plus, minus, and exchange terms, to form a covalent bond. He then called up his associate Fritz London and they worked out the details of the theory over the course of the night. Later, Linus Pauling used the pair bonding ideas of Lewis together with Heitler–London theory to develop two other key concepts in VB theory: resonance (1928) and orbital hybridization (1930). According to Charles Coulson, author of the noted 1952 book *Valence*, this period marks the start of "modern valence bond theory", as contrasted with older valence bond theories, which are essentially electronic theories of valence couched in pre-wave-mechanical terms. Resonance theory was criticized as imperfect by Soviet chemists during the 1950s.

THEORY

According to this theory a covalent bond is formed between the two atoms by the overlap of half filled valence atomic orbitals of each atom containing one unpaired electron. A valence bond structure is similar to a Lewis structure, but where a single Lewis structure cannot be written, several valence bond structures are used. Each of these VB structures represents a specific Lewis structure. This combination of valence bond structures is the main point of resonance theory. Valence bond theory considers that the overlapping atomic orbitals of the participating atoms form a chemical bond. Because of the overlapping, it is most probable that electrons should be in the bond region. Valence bond theory views bonds as weakly coupled orbitals (small overlap). Valence bond theory is typically easier to employ in ground state molecules. The inner-shell orbitals and electrons remain essentially unchanged during the formation of bonds.

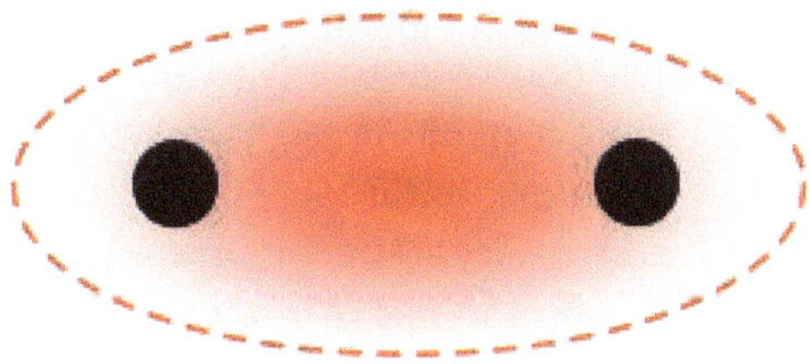

σ bond between two atoms: localization of electron density.

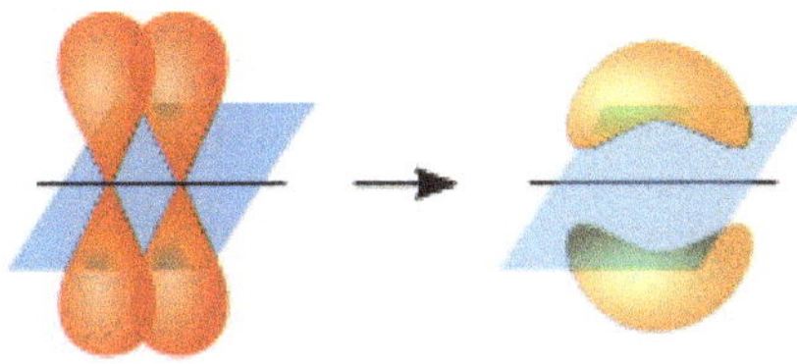

Two p-orbitals forming a π-bond.

The overlapping atomic orbitals can differ. The two types of overlapping orbitals are sigma and pi. Sigma bonds occur when the orbitals of two shared electrons overlap head-to-head. Pi bonds occur when two orbitals overlap when they are parallel. For example, a bond between two *s*-orbital electrons is a sigma bond, because two spheres are always coaxial. In terms of bond order, single bonds have one sigma bond, double bonds consist of one sigma bond and one pi bond, and triple bonds contain one sigma bond and two pi bonds. However, the atomic orbitals for bonding may be hybrids. Often, the bonding atomic orbitals have a character of several possible types of orbitals. The methods to get an atomic orbital with the proper character for the bonding is called hybridization.

VALENCE BOND THEORY TODAY

Modern valence bond theory now complements molecular orbital theory, which does not adhere to the valence bond idea that electron pairs are localized between two specific atoms in a molecule but that they are distributed in sets of molecular orbitals which can extend over the entire molecule. Molecular orbital theory can predict magnetic and ionization properties in a straightforward manner, while valence bond theory gives similar results but is more complicated. Modern valence bond theory views aromatic properties of molecules as due to spin coupling of the π orbitals. This is essentially still the old idea of resonance between Kekulé and Dewar structures. In contrast, molecular orbital theory views aromaticity as delocalization of the π-electrons. Valence bond treatments are restricted to relatively small molecules, largely due to the lack of orthogonality between valence bond orbitals and between valence bond structures, while molecular orbitals are orthogonal. On the other hand, valence bond theory provides a much more accurate picture of the reorganization of electronic charge that takes place when bonds are broken and formed during the course of a chemical reaction. In particular, valence bond theory correctly predicts the dissociation of homonuclear diatomic molecules into separate atoms, while simple molecular orbital theory predicts dissociation into a mixture of atoms and ions. For example, the molecular orbital function for dihydrogen is an equal mixture of the covalent and ionic valence bond structures and so predicts incorrectly that the molecule would dissociate into an equal mixture of hydrogen atoms and hydrogen positive and negative ions.

Modern valence bond theory replaces the overlapping atomic orbitals by overlapping valence bond orbitals that are expanded over a large number of basis functions, either centered each on one atom to give a classical valence bond picture, or centered on all atoms in the molecule. The resulting energies are more competitive with energies from calculations where electron correlation is introduced based on a Hartree–Fock reference wavefunction. The most recent text is by Shaik and Hiberty.

APPLICATIONS OF VALENCE BOND THEORY

An important aspect of the VB theory is the condition of maximum overlap, which leads to the formation of the strongest possible bonds. This theory is used to explain the covalent bond formation in many molecules.

For example, in the case of the F_2 molecule, the F−F bond is formed by the overlap of p_z orbitals of the two F atoms, each containing an unpaired electron. Since the nature of the overlapping orbitals are different in H_2 and F_2 molecules, the bond strength and bond lengths differ between H_2 and F_2molecules.

In an HF molecule the covalent bond is formed by the overlap of the 1*s* orbital of H and the $2p_z$ orbital of F, each containing an unpaired electron. Mutual sharing of electrons between H and F results in a covalent bond in HF.

GENERALIZED VALENCE BOND

The **generalized valence bond** (GVB) method is one of the simplest and oldest valence bond method that uses flexible orbitals in the general way used by modern valence bond theory. The method was developed by the group of William A. Goddard, III around 1970.

Theory

The generalized Coulson–Fischer theory for the hydrogen molecule, discussed in Modern valence bond theory, is used to describe every electron pair in a molecule. The orbitals for each electron pair are expanded in terms of the full basis set and are non-orthogonal. Orbitals from different pairs are forced to be orthogonal - the strong orthogonality condition. This condition simplifies the calculation but can lead to some difficulties.

Calculations

GVB code in some programs, particularly GAMESS (US), can also be used to do a variety of restricted open-shell Hartree–Fock calculations, such as those with one or three electrons in two pi-electron molecular orbitals while retaining the degeneracy of the orbitals. This wave function is essentially a two-determinant function, rather than the one-determinant function of the restricted Hartree–Fock method.

MODERN VALENCE BOND THEORY

Modern valence bond theory is the application of valence bond theory, with computer programs that are competitive in accuracy and economy with programs for the Hartree–Fock method and other molecular orbital based methods. The latter methods dominated quantum chemistry from the advent of digital computers because they were easier to program. The early popularity of valence bond methods thus declined. It is only recently that the programming of valence bond methods has improved. These developments are due to and described by Gerratt, Cooper, Karadakov and Raimondi (1997); Li and McWeeny (2002); Joop H. van Lenthe and co-workers (2002); Song, Mo, Zhang and Wu (2005); and Shaik and Hiberty (2004).

In its simplest form the overlapping atomic orbitals are replaced by orbitals which are expanded as linear combinations of the atom-based basis functions, forming linear combinations of atomic orbitals (LCAO). This expansion is optimized to give the lowest energy. This procedure gives good energies without including ionic structures.

For example, in the hydrogen molecule, classic valence bond theory uses two 1s atomic orbitals (a and b) on the two hydrogen atoms respectively and then constructs a covalent structure:-

$$\Phi_C = (a(1)b(2) + b(1)a(2))\,(\alpha(1)\beta(2) - \beta(1)\alpha(2))$$

and then an ionic structure:-

$$\Phi_I = (a(1)a(2) + b(1)b(2))\,(\alpha(1)\beta(2) - \beta(1)\alpha(2))$$

The final wave function is a linear combination of these two functions. Coulson and Fischer pointed out that a completely equivalent function is:-

$$\Phi_{CF} = ((a+kb)(1)(b+ka)(2) + (b+ka)(1)(a+kb)(2))\,(\alpha(1)\beta(2) - \beta(1)\alpha(2))$$

as expanding this out gives a linear combination of the covalent and ionic structures. Modern valence bond theory replaces the simple linear combination of the two atomic orbitals with a linear combination of all orbitals in a larger basis set. The two resulting valence bond orbitals look like an atomic orbital on one hydrogen atom slightly distorted towards the other hydrogen atom. Modern valence bond theory is thus an extension of this Coulson–Fischer method.

Spin-Coupled Theory

There are a large number of different valence bond methods. Most use n valence bond orbitals for n electrons. If a single set of these orbitals is combined with all linear independent combinations of the spin functions, we have **spin-coupled valence bond theory**. The total wave function is optimized using the variational method by varying the coefficients of the basis functions in the valence bond orbitals and the coefficients of the different spin functions. In other cases only a sub-set of all possible spin functions is used. Many valence bond methods use several sets of the valence bond orbitals. Be warned that different authors use different names for these different valence bond methods.

Valence Bond Programs

Several groups have produced computer programs for modern valence bond calculations that are freely available.

RESONANCE (CHEMISTRY)

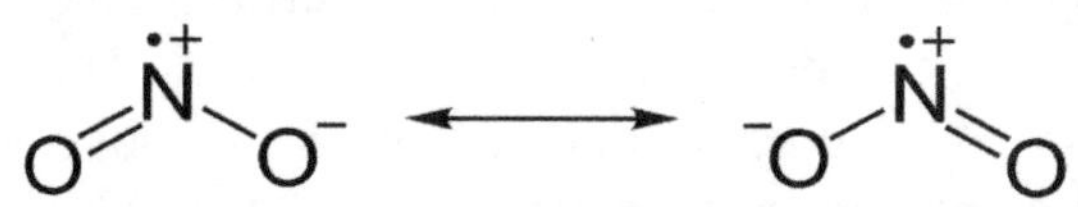

Two of the contributing structures of nitrogen dioxide.

In chemistry, **resonance** or **mesomerism** is a way of describing delocalized electrons within certain molecules or polyatomic ions where the bonding cannot be expressed by one single Lewis structure. A molecule or ion with such delocalized electrons is represented by several **contributing structures**(also called **resonance structures** or **canonical structures**).

Each contributing structure can be represented by a Lewis structure, with only an integer number of covalent bonds between each pair of atoms within the structure. Several Lewis structures are used collectively to describe the actual molecular structure, which is an approximate intermediate between the canonical forms called a **resonance hybrid**. Contributing structures differ only in the position of electrons, not in the position of nuclei.

Electron delocalization lowers the potential energy of the substance and thus makes it more stable than any of the contributing structures. The difference between the potential energy of the actual structure and that of the contributing structure with the lowest potential energy is called the **resonance energy**or delocalization energy.

Resonance is distinguished from isomerization. An isomer is a molecule with the same chemical formula but with different arrangements of atoms in space. Resonance contributors of a molecule, on the contrary, can only differ by the arrangements of electrons. Therefore, the resonance hybrid cannot be represented by a combination of isomers.

The concept first appeared in 1899 in Johannes Thiele›s "Partial Valence Hypothesis" to explain the unusual stability of benzene which would not be expected from August Kekulé›s structure proposed in 1865 with alternating single and double bonds. Benzene undergoes substitution reactions, rather than addition reactions as typical for alkenes. He proposed that the carbon-carbon bond in benzene is intermediate of a single and double bond.

The mechanism of resonance was introduced into quantum mechanics by Werner Heisenberg in 1926 in a discussion of the quantum states of the helium atom. He compared the structure of the helium atom with the classical system of resonating coupled harmonic oscillators. In the classical system, the coupling produces two modes, one of which is lower in frequency than either of the uncoupled vibrations; quantum mechanically, this lower frequency is interpreted as a lower energy. Linus Pauling used this mechanism to explain the partial valence of molecules in 1928, and developed it further in a series of papers in 1931-33. The alternative term *mesomerism* popular in German and French publications with the same meaning was introduced by C. K. Ingold in 1938, but did not catch on in the English literature. The current concept of mesomeric effect has taken on a related but different meaning. The double headed arrow was introduced by the German chemist Fritz Arndt who preferred the German phrase *zwischenstufe* or *intermediate stage*.

In the Soviet Union, resonance theory – especially as developed by Pauling – was attacked in the early 1950s as being contrary to the Marxist principles of dialectical materialism, and in June 1951 the Soviet Academy of Sciences under the leadership of Alexander Nesmeyanov convened a conference on the chemical structure of organic compounds, attended by 400 physicists, chemists, and philosophers, where "the pseudo-scientific essence of the theory of resonance was exposed and unmasked".

General Characteristics of Resonance

Molecules and ions with resonance (also called mesomerism) have the following basic characteristics:

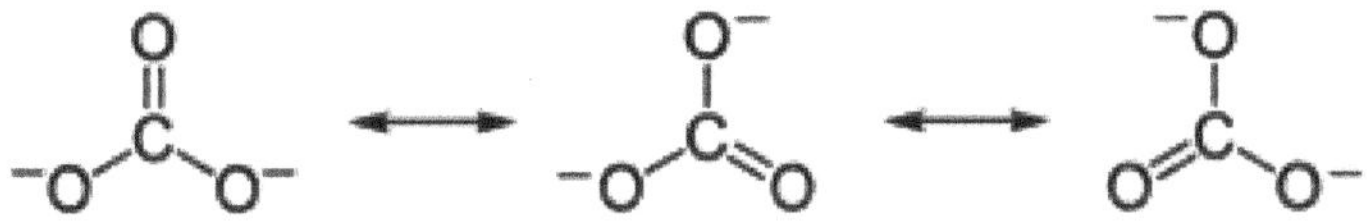

Contributing structures of the carbonate ion

- They can be represented by several correct Lewis formulas, called "contributing structures", "resonance structures" or "canonical forms". The real structure is an intermediate of these structures represented by a resonance hybrid.
- The contributing structures are not isomers. They differ only in the position of electrons, not in the position of nuclei.
- Each Lewis formula must have the same number of valence electrons (and thus the same total charge), and the same number of unpaired electrons, if any.
- Bonds that have different bond orders in different contributing structures do not have typical bond lengths. Measurements reveal intermediate bond lengths.
- The real structure has a lower total potential energy than each of the contributing structures would have. This means that it is more stable than each separate contributing structure would be.

Misconception

It is a common misconception that resonance structures are actual transient states of the molecule, with the molecule oscillating between them or existing as an equilibrium between them. However these individual contributors cannot be observed in the actual resonance-stabilized molecule. Any molecule or ion exists in only one form – the resonance hybrid. Due to confusion with the physical meaning of the word resonance, as no elements actually appear to be resonating, it has been suggested that the term resonance be abandoned in favor of *delocalization*. Resonance energy would thus become *delocalization energy* and a resonance structure becomes a *contributing structure*. The double headed arrows would be replaced by commas to illustrate a set of structures rather than suggesting that there is a reaction that converts among them.

Use of Contributing Structures

In Lewis formulas, electrons are paired between atoms to form covalent bonds. Each single bond is made by two valence electrons, localized between the two bonded atoms. Each double bond has two additional localized π electrons, while each triple bond has four additional π electrons (two pairs) between the bonded atoms.

In molecules or ions that have a combination of one or more single and multiple bonds, often the *exact* position of the respective bonds cannot be indicated by a single Lewis structure. The π electrons appear to be in an intermediate position. To solve this problem, the concept of resonance is used, and the molecule is represented by several contributing structures, each showing a possible distribution of single and multiple bonds. The actual structure has a lowered overall energy and an intermediate bond order.

Resonance Hybrids

The actual structure of a molecule in the *normal quantum state* has the lowest possible value of total energy. This structure is called the "resonance hybrid" of that molecule. The resonance hybrid is the approximate intermediate of the contributing structures, but the overall energy is lower than each of the contributors, due to the resonance energy.

Major and Minor Contributors

One contributing structure may resemble the actual molecule more than another (in the sense of energy and stability). Structures with a low value of potential energy are more stable than those with high values and resemble the actual structure more. The most stable contributing structures are called *major contributors*. Energetically unfavourable and therefore less probable structures are *minor contributors*. Major contributors are generally structures:

- that obey as much as possible the octet rule (8 valence electrons around each atom rather than having deficiencies or surplus)
- that have a maximum number of covalent bonds

- that carry a minimum of charged atoms. If unlike charges are present their separation must be least while for like charges the separation must be maximum.
- with negative charge, if any, on the most electronegative atoms and positive charge, if any, on the most electropositive.

The greater the number of contributing structures, the more stable the molecule. This is because the more states at lower energy are available to the electrons in a particular molecule, the more stable the electrons are. Also the more volume electrons can occupy at lower energy the more stable the molecule is. We can also understand this concept by borrowing a concept of physics. As we know that charge dispersed is directly proportional to stability. Here, electrons can be termed as charged bodies and the more volume they occupy, more the charge gets dispersed ultimately leading to stability.

Equivalent contributors contribute equally to the actual structure; those with low potential energy (the major contributors) contribute more to the resonance hybrid than the less stable minor contributors. Especially when there is more than one major contributor, the resonance stabilization is high. High values of resonance energy are found in aromatic molecules.

Contributing Structures in Diagrams

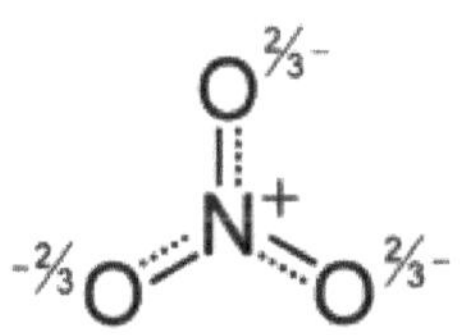

$[S{=}C{=}N^{\ominus} \longleftrightarrow {}^{\ominus}S{-}C{\equiv}N]$

Hybrid of the nitrate ion.

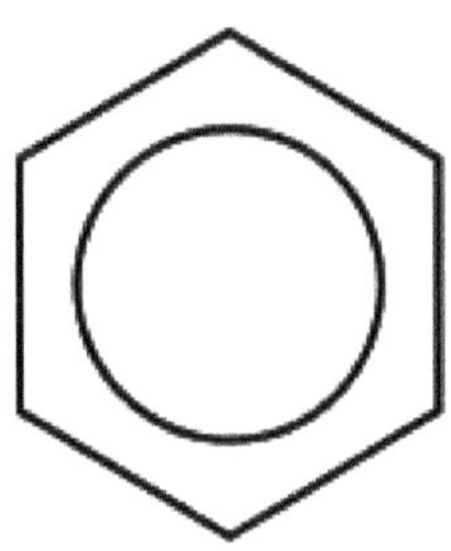

Hybrid of benzene.

In diagrams, contributing structures are typically separated by double-headed arrows (↔). The arrow should not be confused with the right and left pointing *equilibrium arrow* (⇌). All structures together may be enclosed in large square brackets, to indicate they picture one single molecule or ion, not different species in a chemical equilibrium.

Alternatively to the use of resonance structures in diagrams, a hybrid diagram can be used. In a hybrid diagram, pi bond that are involved in resonance are usually pictured as curves or dashed lines, indicating that these are partial rather than normal complete pi bonds. In benzene and other aromatic rings, the delocalized pi-electrons are sometimes pictured as a solid circle.

Examples

Ionic-Covalent Molecules

The ozone molecule is represented by two resonance structures. In reality the two terminal oxygen atoms are equivalent and the hybrid structure is drawn on the right with a charge of $-\frac{1}{2}$ on both oxygen atoms and partial double bonds with a full and dashed line and bond order $1\frac{1}{2}$.

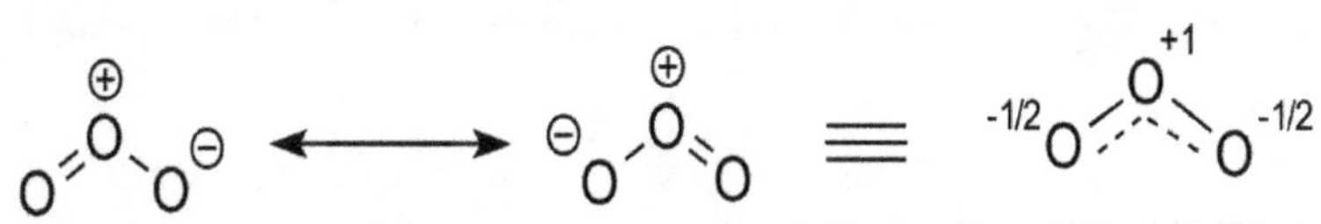

For hypervalent molecules such as xenon difluoride, the rationalization described above can be applied to generate resonance structures to explain the bonding in such molecules. This has been shown by quantum chemical calculations to be the correct description instead of the common expanded octet model.

$[F{-}XeF^{-} \longleftrightarrow F^{-}Xe{-}F]$

Aromatic Molecules

In benzene the two cyclohexatriene *Kekulé* structures, first proposed by Kekulé, are taken together as contributing structures to represent the total structure. In the hybrid structure on the right, the dashed hexagon replaces three double bonds, and represents six electrons in a set of three molecular orbitals of π symmetry, with a nodal plane in the plane of the molecule.

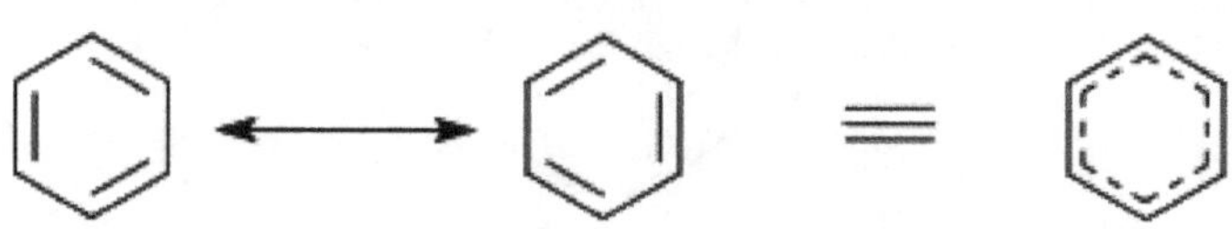

In furan a lone pair of the oxygen atom interacts with the π orbitals of the carbon atoms. The curved arrows depict the permutation of delocalized π electrons, which results in different contributors.

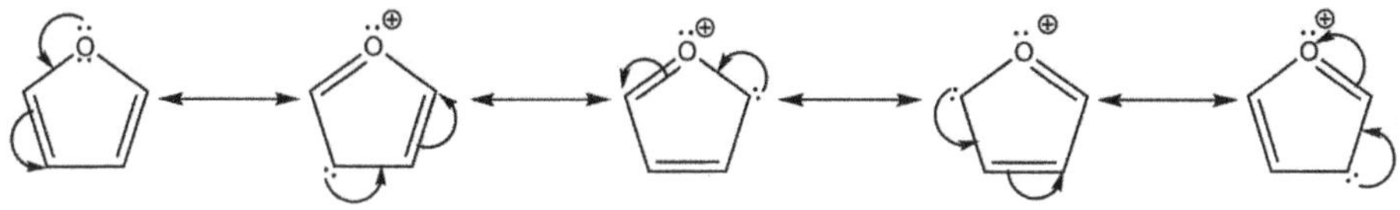

Electron-Deficient Molecules

The diborane molecule is described by resonance structures, each with electron-deficiency on different atoms. This reduces the electron-deficiency on each atom and stabilizes the molecule. Below are the resonance structures of an individual 3c-2e bond in diborane.

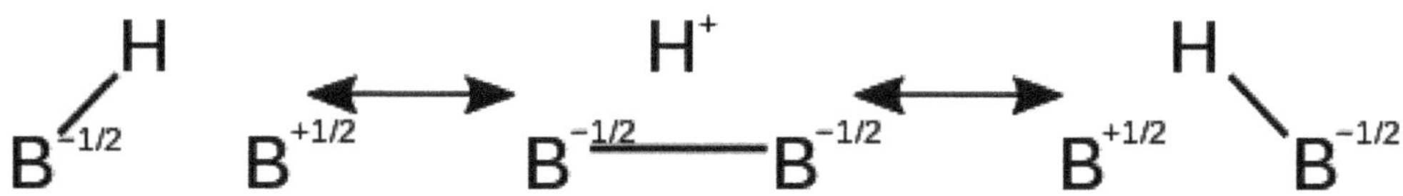

The allyl cation has two contributing structures with a positive charge on the terminal carbon atoms. In the hybrid structure their charge is $+\frac{1}{2}$. The full positive charge can also be depicted as delocalized among three carbon atoms.

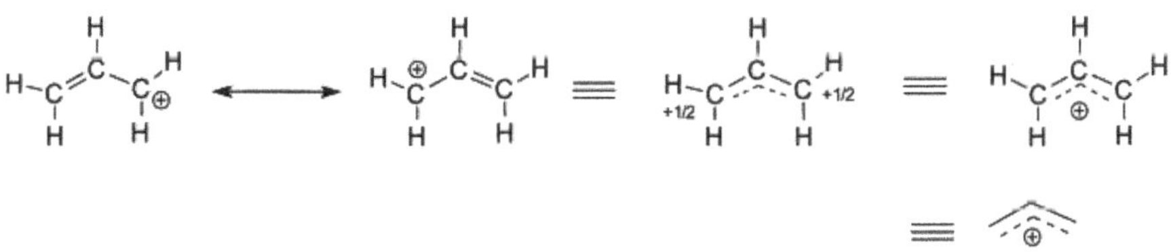

Reactive Intermediates

Often, reactive intermediates such as carbocations and free radicals have more delocalized structure than their parent reactants, giving rise to unexpected products. The classical example is allylic rearrangement. When 1 mole of HCl adds to 1 mole of 1,3-butadiene, in addition to the ordinarily expected product 3-chloro-1-butene, we also find 1-chloro-2-butene. Isotope labelling experiments have shown that what happens here is that the additional double bond shifts from 1,2 position to 2,3 position in some of the product. This and other evidence (such as NMR in superacid solutions) shows that the intermediate carbocation must have a highly delocalized structure, different from its mostly classical (delocalization exists but is small) parent molecule. This cation (an allylic cation) can be represented using resonance, as shown above.

This observation of greater delocalization in less stable molecules is quite general. The excited states of conjugated dienes are stabilised more by conjugation than their ground states, causing them to become organic dyes.

A well-studied example of delocalization that does not involve π electrons (hyperconjugation) can be observed in the non-classical 2-Norbornyl cation. Other examples are diborane (see above) and methanium (CH_5^+). These can be viewed as containing three-center two-electron bonds and are represented either by contributing structures involving rearrangement of σ electrons or by a special notation, a Y that has the three nuclei at its three points.

Delocalized electrons are important for several reasons; a major one is that an expected chemical reaction may not occur because the electrons delocalize to a more stable configuration, resulting in a reaction that happens at a different location. An example is the Friedel–Crafts alkylation of benzene with 1-chloro-2-methylpropane; the carbocation rearranges to a *tert*-butyl group stabilized by hyperconjugation, a particular form of delocalization. Delocalization leads to lengthening of wavelength of electron therefore decreases the energy.

Bond Lengths

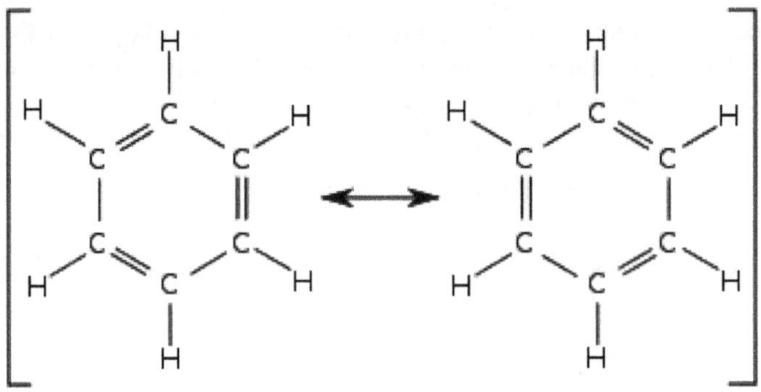

Resonance structures of benzene.

Comparing the two contributing structures of benzene, all single and double bonds are interchanged. Bond lengths can be measured, for example using X-ray diffraction. The average length of a C–C single bond is 154 pm; that of a C=C double bond is 133 pm. In localized cyclohexatriene, the carbon–carbon bonds should be alternating 154 and 133 pm. Instead, all carbon–carbon bonds in benzene are found to be about 139 pm, a bond length intermediate between single and double bond. This mixed single and double bond (or triple bond) character is typical for all molecules in which bonds have a different bond order in different contributing structures. Bond lengths can be compared using bond orders. For example, in cyclohexane the bond order is 1 while that in benzene is 1 + (3 ÷ 6) = 1 $\frac{1}{2}$. Consequently, benzene has more double bond character and hence has a shorter bond length than cyclohexane.

RESONANCE ENERGY

Every structure is associated with a certain quantity of energy, which determines the stability of the molecule or ion (the lower energy, the greater stability). A resonance hybrid has a *structure* that is intermediate between the contributing structures; the total quantity of *potential energy*, however, is lower than the intermediate and the molecule is said to be "stabilized by resonance" or "resonance-stabilized". Hybrids are therefore always more stable than any of the contributing structures would be. The difference between the potential energy of the actual structure (the resonance hybrid) and that of the contributing structure with the lowest potential energy is called the "resonance energy".

Resonance Energy of Benzene

Resonance (or delocalization) energy is the amount of energy needed to convert the true delocalized structure into that of the most stable contributing structure. The *empirical resonance energy* can be estimated by comparing the enthalpy change of hydrogenation of the real substance with that estimated for the contributing structure.

The complete hydrogenation of benzene to cyclohexane via 1,3-cyclohexadiene and cyclohexene is exothermic; 1 mole of benzene delivers 208.4 kJ (49.8 kcal).

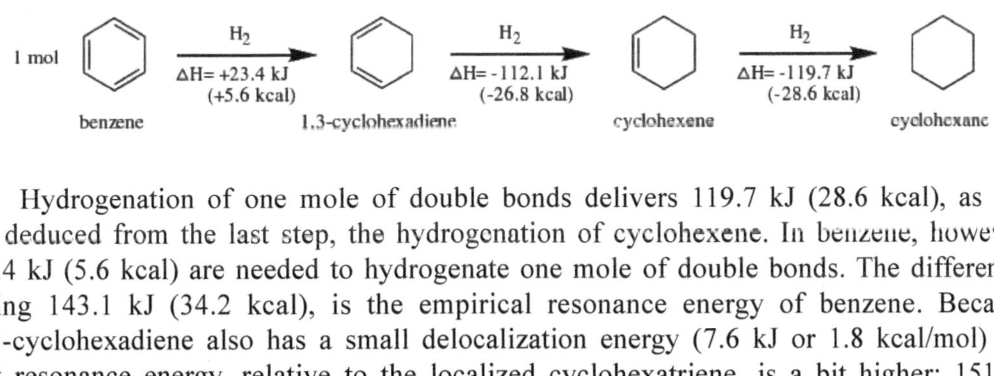

Hydrogenation of one mole of double bonds delivers 119.7 kJ (28.6 kcal), as can be deduced from the last step, the hydrogenation of cyclohexene. In benzene, however, 23.4 kJ (5.6 kcal) are needed to hydrogenate one mole of double bonds. The difference, being 143.1 kJ (34.2 kcal), is the empirical resonance energy of benzene. Because 1,3-cyclohexadiene also has a small delocalization energy (7.6 kJ or 1.8 kcal/mol) the net resonance energy, relative to the localized cyclohexatriene, is a bit higher: 151 kJ or 36 kcal/mol.

This measured resonance energy is also the difference between the hydrogenation energy of three 'non-resonance' double bonds and the measured hydrogenation energy: $(3 \times 119.7) - 208.4 = 150.7$ kJ/mol (36 kcal).

Resonance in Valence Bond (VB) Theory

Resonance has a deeper significance in the mathematical formalism of valence bond theory (VB). Quantum mechanics requires that the wavefunction of a molecule obeys its observed symmetry. If a single contributing structure does not achieve this, resonance is invoked.

For example, in benzene, valence bond theory begins with the two Kekulé structures and constructs the actual wave function of the molecule as a linear superposition of the wave functions representing the two structures. As both Kekulé structures have equal energy, they are equal contributors to the overall structure – the superposition is an equally weighted average, or a 1:1 linear combination of the two – but this need not be the case. The symmetric combination gives the ground state while the antisymmetric combination gives the first excited state as shown.

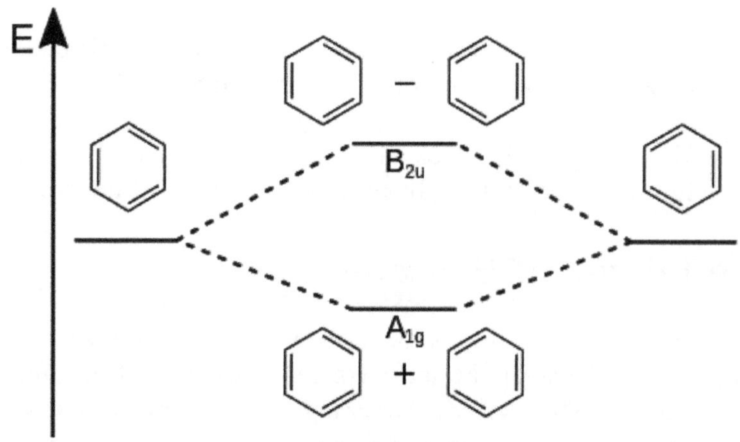

VB mixing diagram of benzene. The A_{1g} and B_{2u} labels define the symmetries of the two states, as defined by the character table for the D_{6h} symmetry group.

In general, the superposition is written with undetermined coefficients, which are then variationally optimized to find the lowest possible energy for the given set of basis wave functions. When more contributing structures are included, the molecular wave function becomes more accurate and more excited states can be derived from different combinations of the contributing structures.

Comparison with Molecular Orbital (MO) Theory

In molecular orbital theory, the main alternative to valence bond theory, the equivalent of the *symmetry-adapted linear combination* role of resonance is the linear combination of atomic orbitals. In MO theory, the molecular orbitals (MOs) are approximated as sums of all the atomic orbitals (AOs) on all the atoms; there are as many MOs as AOs. Each AO_i has a *weighting* coefficient c_i that indicates the AO's contribution to a particular MO. For example, in benzene, the MO model gives us 6 π MOs which are combinations of the $2p_z$ AOs on each of the 6 C atoms. Thus, each π MO is delocalized over the whole benzene molecule and any electron *occupying* an MO will be delocalized over the whole molecule. This MO interpretation has inspired the picture of the benzene ring as a hexagon with a circle inside. When describing benzene, the VB concept of localized sigma ‹bonds› and the MO concept of ‹delocalized› π electrons are frequently combined in elementary chemistry courses.

The resonance structures in the VB model are particularly useful in predicting the effect of substituents on π systems such as benzene. They lead to the models of resonance structures for an electron-withdrawing group and electron-releasing group on benzene. The utility of MO theory is that a quantitative indication of the charge from the π system on an atom can be obtained from the squares of the *weighting* coefficient c_i on atom C_i. Charge $q_i \approx c2$ i. The reason for squaring the coefficient is that if an electron is described by an AO, then the square of the AO gives the electron density. The AOs are adjusted (normalized) so that $AO^2 = 1$, and $q_i \approx (c_iAO_i)^2 \approx c2$ i. In benzene, $q_i = 1$ on each C atom. With an electron-withdrawing group $q_i < 1$ on the *ortho* and *para* C atoms and $q_i > 1$ for an electron-releasing group.

Coefficients

Weighting of the resonance structures in terms of their contribution to the overall structure can be calculated in multiple ways, using *"Ab initio"* methods derived from Valence Bond theory, or else from the Natural Bond Orbitals (NBO) approaches of Weinhold NBO5, or finally from empirical calculations based on the Hückel method. A Hückel method-based software for teaching resonance is available on the HuLiS Web site.

Charge Delocalization

In the case of ions it is common to speak about delocalized charge (charge delocalization). An example of delocalized charge in ions can be found in the carboxylate group, wherein the negative charge is centered equally on the two oxygen atoms. Charge delocalization in anions is an important factor determining their reactivity (generally: the higher the extent of delocalization the lower the reactivity) and, specifically, the acid strength of their conjugate acids. As a general rule, the better delocalized is the charge in an anion the stronger is its conjugate acid. For example, the negative charge in perchlorate anion (ClO_4^-) is evenly distributed among the symmetrically oriented oxygen atoms (and a part of it is also kept by the central chlorine atom). This excellent charge delocalization combined with the high number of oxygen atoms (four) and high electronegativity of the central chlorine atom leads to perchloric acid being one of the strongest known acids with a pK_a value of −10. The extent of charge delocalization in an anion can be quantitatively expressed via the WAPS (weighted average positive sigma) parameter parameter and an analogous WANS (weighted average negative sigma)parameter is used for cations.

WAPS values of anions of common acids and WANS values of cations of common bases			
Compound	**WAPS × 10^5**	**Compound**	**WANS × 10^5**
$(C_2F_5SO_2)_2NH$	2.0	Triphenylphosphine	2.1
$(CF_3)_3COH$	3.6	Phenyl tetramethylguanidine	2.5
Picric acid	4.3	Tripropylamine	2.6
2,4-Dinitrophenol	4.9	MTBD (7-Methyl-triazabicyclodecene)	2.9
Benzoic acid	7.1	DBU (1,8-Diazabicycloundec-7-ene)	3.0
Phenol	8.8	TBD (Triazabicyclodecene)	3.5
Acetic acid	16.1	*N,N*-Dimethylaniline	4.7

HI	21.9	Pyridine	7.2
HBr	29.1	Aniline	8.2
HCl	35.9	Propylamine	8.9

WAPS and WANS values are given in e/Å^4. Larger values indicate more localized charge in the corresponding ion.

RFERENCES

1. IUPAC, Compendium of Chemical Terminology, 2nd ed. (the "Gold Book") (1997). Online corrected version: (2006–) "Mesomerism".
2. IUPAC, Compendium of Chemical Terminology, 2nd ed. (the "Gold Book") (1997). Online corrected version: (2006–) "Resonance".
3. IUPAC, Compendium of Chemical Terminology, 2nd ed. (the "Gold Book") (1997). Online corrected version: (2006–) "contributing structure".
4. Pauling, Linus (1960). "The Concept of Resonance". The Nature of the Chemical Bond – An Introduction to Modern Structural Chemistry *(3rd* ed.). Cornell University Press. pp. 10–13.
5. IUPAC, Compendium of Chemical Terminology, 2nd ed. (the "Gold Book") (1997). Online corrected version: (2006–) "resonance energy".
6. Resonance Forms". UCDavis Chem Wiki. UCDavis. Retrieved 7 October 2015.
7. Pauling, Linus, Resonance, p. 1
8. The Science and Humanism of Linus Pauling". See last paragraph of section 1.
9. Pauling, L. (1960). The Nature of the Chemical Bond (3rd ed.). Oxford University Press. p. 184. In this source, Pauling first mentions related papers by Slater and Hückel in 1931, and then cites his own key papers: J. Amer. Chem. Soc. **53** *(1367): 3225. 1931.* Missing or empty |title= (help) and subsequent papers in 1932–33.
10. Moore, Barrington, Jr. *(1954). Terror and Progress USSR: Some Sources of Change and Stability in the Soviet Dictatorship. pp. 142–143.*
11. Pauling, Linus (July 29, 1946), Resonance *(Manuscript for publication in* Encyclopædia Britannica), p. 13
12. Kerber, Robert C. (2006). "If It›s Resonance, What Is Resonating?". *J. Chem. Educ.* **83**: 223. doi:10.1021/ed083p223.
13. Graphical representation for chemical structure diagrams", IUPAC Recommendations 2008, IUPAC, p. 387 (GR-8)
14. Graphical representation for chemical structure diagrams", IUPAC Recommendations 2008, IUPAC, pp. 379–382 (GR–6)
15. Wade, G. Organic Chemistry (6th ed.).
16. Bruice, Paula Y. Organic Chemistry (4th ed.).

17. Morrison, Robert; Boyd, Robert (1989). "Chapter 10". Organic Chemistry (5th ed.). Prentice Hall of India. p. 372. ISBN 0-87692-560-3. The resonance hybrid is more stable than any of the contributing structures.

18. Wiberg; Nakaji; Morgan (1993). "Heat of hydrogenation of a cis imine. An experimental and theoretical study". J. Am. Chem. Soc. **115** *(9): 3527–3532.* doi:10.1021/ja00062a017.

19. Sherman, J. (February 1939). "The heats of hydrogenation of unsaturated hydrocarbons". J. Am. Oil Chem. Soc. **16** *(2): 28.* doi:10.1007/BF02543208.

20. Shaik, Sason S.; Hiberty, Phillipe C. (2008). A Chemist›s Guide to Valence Bond Theory. New Jersey: Wiley-Interscience. pp. 200–203. ISBN 978-0-470-03735-5.

21. Sellers, Kathleen; Weeks, Katherine; Alsop, William R.; Clough, Stephen R.; Hoyt, Marilyn; Pugh, Barbara (2006). Perchlorate: Environmental problems and solutions. CRC Press. p. 16. ISBN 0-8493-8081-2.

22. Kaupmees, K.; Kaljurand, I.; Leito, I. (2010). "Influence of Water Content on the Acidities in Acetonitrile. Quantifying Charge Delocalization in Anions". J. Phys. Chem. A. **114**: 11788–11793. doi:10.1021/jp105670t.

23. Kaupmees, K.; Kaljurand, I.; Leito, I. (2014). "Influence of Water Content on Basicities in Acetonitrile". J. Sol. Chem. **43**: 1270–1281. doi:10.1007/s10953-014-0201-4.

24. Kaupmees, K.; Trummal,, A.; Leito, I. (2014). "Basicities of Strong Bases in Water: A Computational Study". Croat. Chem. Acta. **87**: 385–395. doi:10.5562/cca2472.

25. Raamat, E.; Kaupmees, K.; Ovsjannikov, G.; Trummal, A.; Kütt, A.; Saame, J.; Koppel, I.; Kaljurand, I.; Lipping, L.; Rodima, T.; Pihl, V.; Koppel, I. A.; Leito, I. (2013). "Acidities of strong neutral Brønsted acids in different media". J. Phys. Org. Chem. **26**: 162–170. doi:10.1002/poc.2946.

CHAPTER 4

Atomic Orbitals

Chapter Contents

- Introducton
- Electron Properties
- History
- Orbital Names
- Hydrogen-Like Orbitals
- Quantum Numbers
- Shapes of Orbitals
- Orbital Energy
- Transitions Between Orbitals
- Atoms in Molecules

When a planet moves around the sun, you can plot a definite path for it which is called an orbit. A simple view of the atom looks similar and you may have pictured the electrons as orbiting around the nucleus. The truth is different, and electrons in fact inhabit regions of space known as ***orbitals.***

Orbits and orbitals sound similar, but they have quite different meanings. It is essential that you understand the difference between them.

INTRODUCTON

In quantum mechanics, an **atomic orbital** is a mathematical function that describes the wave-like behavior of either one electron or a pair of electrons in an atom. This function can be used to calculate the probability of finding any electron of an atom in any specific region around the atom's nucleus. The term, atomic orbital, may also refer to the physical region or space where the electron can be calculated to be present, as defined by the particular mathematical form of the orbital.

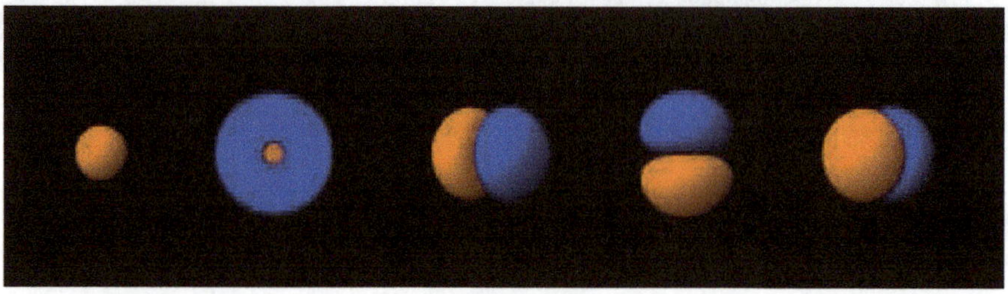

The shapes of the first five atomic orbitals are: 1s, 2s, $2p_x$, $2p_y$, and $2p_z$. The two colors show the phase or sign of the wave function in each region. These are graphs of $\psi(x,y,z)$ functions which depend on the coordinates of one electron. To see the elongated shape of $\psi(x,y,z)^2$ functions that show probability density more directly, see the graphs of d-orbitals below.

Each orbital in an atom is characterized by a unique set of values of the three quantum numbers *n*, *l*, and *m*, which respectively correspond to the electron's energy, angular momentum, and an angular momentum vector component (the magnetic quantum number). Each such orbital can be occupied by a maximum of two electrons, each with its own spin quantum number *s*. The simple names **s orbital**, **p orbital**, **d orbital** and **f orbital** refer to orbitals with angular momentum quantum number $l = 0$, 1, 2 and 3 respectively. These names, together with the value of *n*, are used to describe the electron configurations of atoms. They are derived from the description by early spectroscopists of certain series of alkali metal spectroscopic lines as **s**harp, **p**rincipal, **d**iffuse, and **f**undamental. Orbitals for $l > 3$ continue alphabetically, omitting j (g, h, i, k, …) because some languages do not distinguish between the letters "i" and "j".

Atomic orbitals are the basic building blocks of the **atomic orbital model** (alternatively known as the electron cloud or wave mechanics model), a modern framework for visualizing the submicroscopic behavior of electrons in matter. In this model the electron cloud of a multi-electron atom may be seen as being built up (in approximation) in an electron configuration that is a product of simpler hydrogen-like atomic orbitals. The repeating *periodicity* of the blocks of 2, 6, 10, and 14 elements within sections of the periodic table arises naturally from the total number of electrons that occupy a complete set of **s**, **p**, **d** and **f** atomic orbitals, respectively, although for higher values of the quantum number *n*, particularly when the atom in question bears a positive charge, the energies of certain sub-shells become very similar and so the order in which they are said to be populated by electrons (e.g. Cr = $[Ar]4s^13d^5$ and $Cr^{2+} = [Ar]3d^4$) can only be rationalized somewhat arbitrarily.

ELECTRON PROPERTIES

With the development of quantum mechanics and experimental findings (such as the two slits diffraction of electrons), it was found that the orbiting electrons around a nucleus could not be fully described as particles, but needed to be explained by the wave-particle duality. In this sense, the electrons have the following properties:

Wave-like properties:

- The electrons do not orbit the nucleus in the manner of a planet orbiting the sun, but instead exist as standing waves. The lowest possible energy an electron can take is therefore analogous to the fundamental frequency of a wave on a string. Higher energy states are then similar to harmonics of the fundamental frequency.
- The electrons are never in a single point location, although the probability of interacting with the electron at a single point can be found from the wave function of the electron.

Particle-like properties:

- There is always an integer number of electrons orbiting the nucleus.
- Electrons jump between orbitals in a particle-like fashion. For example, if a single photon strikes the electrons, only a single electron changes states in response to the photon.
- The electrons retain particle-like properties such as: each wave state has the same electrical charge as the electron particle. Each wave state has a single discrete spin (spin up or spin down). This can depend upon its superposition.

Thus, despite the popular analogy to planets revolving around the Sun, electrons cannot be described simply as solid particles. In addition, atomic orbitals do not closely resemble a planet's elliptical path in ordinary atoms. A more accurate analogy might be that of a large and often oddly shaped "atmosphere" (the electron), distributed around a relatively tiny planet (the atomic nucleus). Atomic orbitals exactly describe the shape of this "atmosphere" only when a single electron is present in an atom. When more electrons are added to a single atom, the additional electrons tend to more evenly fill in a volume of space around the nucleus so that the resulting collection (sometimes termed the atom's "electron cloud") tends toward a generally spherical zone of probability describing where the atom's electrons will be found. This is due to the uncertainty principle.

Formal Quantum Mechanical Definition

Atomic orbitals may be defined more precisely in formal quantum mechanical language. Specifically, in quantum mechanics, the state of an atom, i.e., an eigenstate of the atomic Hamiltonian, is approximated by an expansion (see configuration interaction expansion and basis set) into linear combinations of anti-symmetrized products (Slater determinants) of one-electron functions. The spatial components of these one-electron functions are called atomic orbitals. (When one considers also their spin component, one speaks of **atomic spin orbitals**.) A state is actually a function of the coordinates of all the electrons, so that their motion is correlated, but this is often approximated by this independent-particle model of products of single electron wave functions. (The London dispersion force, for example, depends on the correlations of the motion of the electrons.)

In atomic physics, the atomic spectral lines correspond to transitions (quantum leaps) between quantum states of an atom. These states are labeled by a set of quantum numbers summarized in the term symbol and usually associated with particular electron configurations, i.e., by occupation schemes of atomic orbitals (for example, $1s^2\ 2s^2\ 2p^6$ for the ground state of neon—term symbol: $^1S_0$).

This notation means that the corresponding Slater determinants have a clear higher weight in the configuration interaction expansion. The atomic orbital concept is therefore a key concept for visualizing the excitation process associated with a given transition. For example, one can say for a given transition that it corresponds to the excitation of an electron from an occupied orbital to a given unoccupied orbital. Nevertheless, one has to keep in mind that electrons are fermions ruled by the Pauli Exclusion Principle and cannot be distinguished from the other electrons in the atom. Moreover, it sometimes happens that the configuration interaction expansion converges very slowly and that one cannot speak about simple one-determinant wave function at all. This is the case when electron correlation is large.

Fundamentally, an atomic orbital is a one-electron wave function, even though most electrons do not exist in one-electron atoms, and so the one-electron view is an approximation. When thinking about orbitals, we are often given an orbital vision which (even if it is not spelled out) is heavily influenced by this Hartree–Fock approximation, which is one way to reduce the complexities of molecular orbital theory.

Types of Orbitals

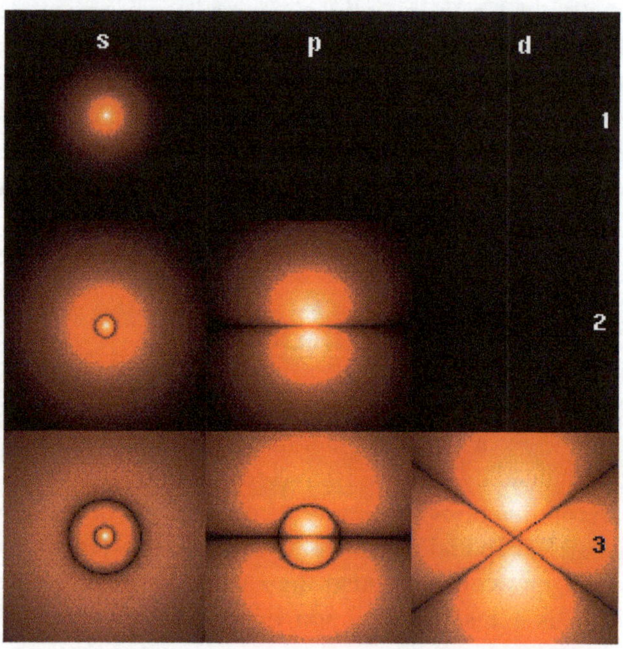

False-color density images of some hydrogen-like atomic orbitals (**f** orbitals and higher are not shown).

Atomic orbitals can be the hydrogen-like "orbitals" which are exact solutions to the Schrödinger equation for a hydrogen-like "atom" (i.e., an atom with one electron). Alternatively, atomic orbitals refer to functions that depend on the coordinates of one electron (i.e., orbitals) but are used as starting points for approximating wave functions that depend on the simultaneous coordinates of all the electrons in an atom or molecule. The coordinate systems chosen for atomic orbitals are usually spherical coordinates (r, θ, φ) in atoms and cartesians (x, y, z) in polyatomic molecules. The advantage of spherical coordinates (for atoms) is that an orbital wave function is a product of three factors each dependent on a single coordinate: $\psi(r, \theta, \varphi) = R(r)\Theta(\theta)\Phi(\varphi)$. The angular factors of atomic orbitals $\Theta(\theta)\Phi(\varphi)$ generate s, p, d, etc. functions as real combinations of spherical harmonics $Y_{lm}(\theta, \varphi)$ (where l and m are quantum numbers). There are typically three mathematical forms for the radial functions $R(r)$ which can be chosen as a starting point for the calculation of the properties of atoms and molecules with many electrons:

- The *hydrogen-like atomic orbitals* are derived from the exact solution of the Schrödinger Equation for one electron and a nucleus, for a hydrogen-like atom. The part of the function that depends on the distance from the nucleus has nodes (radial nodes) and decays as $e^{-(\text{constant} \times \text{distance})}$.
- The Slater-type orbital (STO) is a form without radial nodes but decays from the nucleus as does the hydrogen-like orbital.
- The form of the Gaussian type orbital (Gaussians) has no radial nodes and decays as $e^{(-\text{distance squared})}$.

Although hydrogen-like orbitals are still used as pedagogical tools, the advent of computers has made STOs preferable for atoms and diatomic molecules since combinations of STOs can replace the nodes in hydrogen-like atomic orbital. Gaussians are typically used in molecules with three or more atoms. Although not as accurate by themselves as STOs, combinations of many Gaussians can attain the accuracy of hydrogen-like orbitals.

HISTORY

The term "orbital" was coined by Robert Mulliken in 1932 as an abbreviation for *one-electron orbital wave function*. However, the idea that electrons might revolve around a compact nucleus with definite angular momentum was convincingly argued at least 19 years earlier by Niels Bohr, and the Japanese physicist Hantaro Nagaoka published an orbit-based hypothesis for electronic behavior as early as 1904. Explaining the behavior of these electron "orbits" was one of the driving forces behind the development of quantum mechanics.

Early Models

With J.J. Thomson's discovery of the electron in 1897, it became clear that atoms were not the smallest building blocks of nature, but were rather composite particles. The newly discovered structure within atoms tempted many to imagine how the atom's constituent parts might interact with each other. Thomson theorized that multiple electrons revolved in orbit-like rings within a positively charged jelly-like substance, and between the electron's discovery and 1909, this "plum pudding model" was the most widely accepted explanation of atomic structure.

Shortly after Thomson's discovery, Hantaro Nagaoka predicted a different model for electronic structure. Unlike the plum pudding model, the positive charge in Nagaoka's "Saturnian Model" was concentrated into a central core, pulling the electrons into circular orbits reminiscent of Saturn's rings. Few people took notice of Nagaoka's work at the time, and Nagaoka himself recognized a fundamental defect in the theory even at its conception, namely that a classical charged object cannot sustain orbital motion because it is accelerating and therefore loses energy due to electromagnetic radiation. Nevertheless, the Saturnian model turned out to have more in common with modern theory than any of its contemporaries.

Bohr Atom

In 1909, Ernest Rutherford discovered that the bulk of the atomic mass was tightly condensed into a nucleus, which was also found to be positively charged. It became clear from his analysis in 1911 that the plum pudding model could not explain atomic structure. In 1913 as Rutherford's post-doctoral student, Niels Bohr proposed a new model of the atom, wherein electrons orbited the nucleus with classical periods, but were only permitted to have discrete values of angular momentum, quantized in units $h/2\pi$. This constraint automatically permitted only certain values of electron energies. The Bohr model of the atom fixed the problem of energy loss from radiation from a ground state (by declaring that there was no state below this), and more importantly explained the origin of spectral lines.

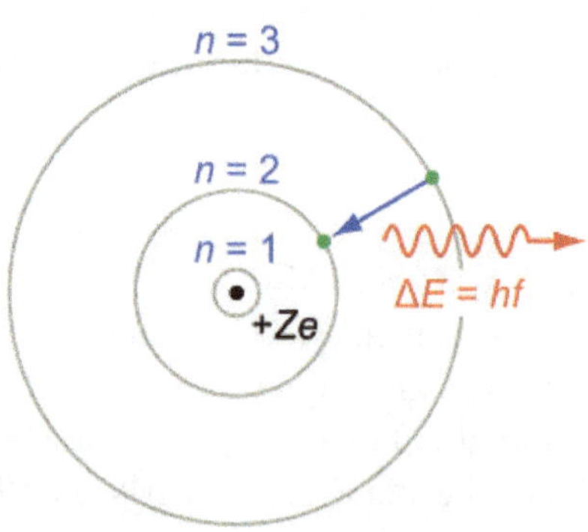

The Rutherford–Bohr model of the hydrogen atom.

After Bohr's use of Einstein's explanation of the photoelectric effect to relate energy levels in atoms with the wavelength of emitted light, the connection between the structure of electrons in atoms and the emission and absorption spectra of atoms became an increasingly useful tool in the understanding of electrons in atoms. The most prominent feature of emission and absorption spectra (known experimentally since the middle of the 19th century), was that these atomic spectra contained discrete lines. The significance of the Bohr model was that it related the lines in emission and absorption spectra to the energy differences between the orbits that electrons could take around an atom. This was, however, *not* achieved by Bohr through giving the electrons some kind of wave-like properties, since the idea that electrons could behave as matter waves was not suggested until eleven years later. Still, the Bohr model's use of quantized angular momenta and therefore quantized energy levels was a significant step towards the understanding of electrons in atoms, and also a significant step towards the development of quantum mechanics in suggesting that quantized restraints must account for all discontinuous energy levels and spectra in atoms.

With de Broglie's suggestion of the existence of electron matter waves in 1924, and for a short time before the full 1926 Schrödinger equation treatment of hydrogen-like atom, a Bohr electron "wavelength" could be seen to be a function of its momentum, and thus a Bohr orbiting electron was seen to orbit in a circle at a multiple of its half-wavelength (this physically incorrect Bohr model is still often taught to beginning students). The Bohr model for a short time could be seen as a classical model with an additional constraint provided by the ‹wavelength' argument. However, this period was immediately superseded by the full three-dimensional wave mechanics of 1926. In our current understanding of physics, the Bohr model is called a semi-classical model because of its quantization of angular momentum, not primarily because of its relationship with electron wavelength, which appeared in hindsight a dozen years after the Bohr model was proposed.

The Bohr model was able to explain the emission and absorption spectra of hydrogen. The energies of electrons in the $n = 1, 2, 3$, etc. states in the Bohr model match those of current physics. However, this did not explain similarities between different atoms, as expressed by the periodic table, such as the fact that helium (two electrons), neon (10 electrons), and argon (18 electrons) exhibit similar chemical inertness. Modern quantum mechanics explains this in terms of electron shells and subshells which can each hold a number of electrons determined by the Pauli exclusion principle. Thus the $n = 1$ state can hold one or two electrons, while the $n = 2$ state can hold up to eight electrons in 2s and 2p subshells. In helium, all $n = 1$ states are fully occupied; the same for $n = 1$ and $n = 2$ in neon. In argon the 3s and 3p subshells are similarly fully occupied by eight electrons; quantum mechanics also allows a 3d subshell but this is at higher energy than the 3s and 3p in argon (contrary to the situation in the hydrogen atom) and remains empty.

Modern Conceptions and Connections to the Heisenberg Uncertainty Principle

Immediately after Heisenberg discovered his uncertainty principle, it was noted by Bohr that the existence of any sort of wave packet implies uncertainty in the wave frequency and wavelength, since a spread of frequencies is needed to create the packet itself. In quantum mechanics, where all particle momenta are associated with waves, it is the formation of such a wave packet which localizes the wave, and thus the particle, in space. In states where a quantum mechanical particle is bound, it must be localized as a wave packet, and the existence of the packet and its minimum size implies a spread and minimal value in particle wavelength, and thus also momentum and energy. In quantum mechanics, as a particle is localized to a smaller region in space, the associated compressed wave packet requires a larger and larger range of momenta, and thus larger kinetic energy. Thus, the binding energy to contain or trap a particle in a smaller region of space, increases without bound, as the region of space grows smaller. Particles cannot be restricted to a geometric point in space, since this would require an infinite particle momentum.

In chemistry, Schrödinger, Pauling, Mulliken and others noted that the consequence of Heisenberg's relation was that the electron, as a wave packet, could not be considered to have an exact location in its orbital. Max Born suggested that the electron's position needed to be described by a probability distribution which was connected with finding the electron at some point in the wave-function which described its associated wave packet. The new quantum mechanics did not give exact results, but only the probabilities for the occurrence of a variety of possible such results. Heisenberg held that the path of a moving particle has no meaning if we cannot observe it, as we cannot with electrons in an atom.

In the quantum picture of Heisenberg, Schrödinger and others, the Bohr atom number *n* for each orbital became known as an *n-sphere* in a three dimensional atom and was pictured as the mean energy of the probability cloud of the electron's wave packet which surrounded the atom.

ORBITAL NAMES

Orbitals are given names in the form:

$$X\,\mathbf{type}^{y}$$

where *X* is the energy level corresponding to the principal quantum number *n*, **type** is a lower-case letter denoting the shape or subshell of the orbital and it corresponds to the angular quantum number *l*, and *y*is the number of electrons in that orbital.

For example, the orbital $1s^2$ (pronounced "one ess two") has two electrons and is the lowest energy level ($n = 1$) and has an angular quantum number of $l = 0$. In X-ray notation, the principal quantum number is given a letter associated with it. For $n = 1, 2, 3, 4, 5, \ldots$, the letters associated with those numbers are K, L, M, N, O, ... respectively.

HYDROGEN-LIKE ORBITALS

The simplest atomic orbitals are those that are calculated for systems with a single electron, such as the hydrogen atom. An atom of any other element ionized down to a single electron is very similar to hydrogen, and the orbitals take the same form. In the Schrödinger equation for this system of one negative and one positive particle, the atomic orbitals are the eigenstates of the Hamiltonian operator for the energy. They can be obtained analytically, meaning that the resulting orbitals are products of a polynomial series, and exponential and trigonometric functions. (see hydrogen atom).

For atoms with two or more electrons, the governing equations can only be solved with the use of methods of iterative approximation. Orbitals of multi-electron atoms are *qualitatively* similar to those of hydrogen, and in the simplest models, they are taken to have the same form. For more rigorous and precise analysis, the numerical approximations must be used.

A given (hydrogen-like) atomic orbital is identified by unique values of three quantum numbers: *n*, *l*, and m_l. The rules restricting the values of the quantum numbers, and their energies (see below), explain the electron configuration of the atoms and the periodic table.

The stationary states (quantum states) of the hydrogen-like atoms are its atomic orbitals. However, in general, an electron's behavior is not fully described by a single orbital. Electron states are best represented by time-depending "mixtures" (linear combinations) of multiple orbitals. See Linear combination of atomic orbitals molecular orbital method.

The quantum number *n* first appeared in the Bohr model where it determines the radius of each circular electron orbit. In modern quantum mechanics however, *n* determines the mean distance of the electron from the nucleus; all electrons with the same value of *n* lie at the same average distance. For this reason, orbitals with the same value of *n* are said to comprise a "shell". Orbitals with the same value of *n* and also the same value of *l* are even more closely related, and are said to comprise a "subshell".

QUANTUM NUMBERS

Because of the quantum mechanical nature of the electrons around a nucleus, atomic orbitals can be uniquely defined by a set of integers known as quantum numbers. These quantum numbers only occur in certain combinations of values, and their physical interpretation changes depending on whether real or complex versions of the atomic orbitals are employed.

Complex Orbitals

In physics, the most common orbital descriptions are based on the solutions to the hydrogen atom, where orbitals are given by the product between a radial function and a pure spherical harmonic. The quantum numbers, together with the rules governing their possible values, are as follows:

The principal quantum number *n* describes the energy of the electron and is always a positive integer. In fact, it can be any positive integer, but for reasons discussed below, large numbers are seldom encountered. Each atom has, in general, many orbitals associated with each value of *n*; these orbitals together are sometimes called *electron shells*.

The azimuthal quantum number *l* describes the orbital angular momentum of each electron and is a non-negative integer. Within a shell where *n* is some integer n_0, *l* ranges across all (integer) values satisfying the relation $0 \leq \ell \leq n_0 - 1$. For instance, the $n = 1$ shell has only orbitals with $\ell = 0$, and the $n = 2$ shell has only orbitals with $\ell = 0$, and $\ell = 1$. The set of orbitals associated with a particular value of *l* are sometimes collectively called a *subshell*.

The magnetic quantum number, m_ℓ, describes the magnetic moment of an electron in an arbitrary direction, and is also always an integer. Within a subshell where ℓ is some integer ℓ_0, m_ℓ ranges thus: $-\ell_0 \leq m_\ell \leq \ell_0$.

The above results may be summarized in the following table. Each cell represents a subshell, and lists the values of m_ℓ available in that subshell. Empty cells represent subshells that do not exist.

	$\ell = 0$	$\ell = 1$	$\ell = 2$	$\ell = 3$	$\ell = 4$	...
$n = 1$	$m_\ell = 0$					
$n = 2$	0	−1, 0, 1				
$n = 3$	0	−1, 0, 1	−2, −1, 0, 1, 2			
$n = 4$	0	−1, 0, 1	−2, −1, 0, 1, 2	−3, −2, −1, 0, 1, 2, 3		
$n = 5$	0	−1, 0, 1	−2, −1, 0, 1, 2	−3, −2, −1, 0, 1, 2, 3	−4, −3, −2, −1, 0, 1, 2, 3, 4	
...	...	...	...	...	...	...

Subshells are usually identified by their n- and ℓ-values. n is represented by its numerical value, but ℓ is represented by a letter as follows: 0 is represented by 's', 1 by 'p', 2 by 'd', 3 by 'f', and 4 by 'g'. For instance, one may speak of the subshell with $n = 2$ and $\ell = 0$ as a '2s subshell'.

Each electron also has a spin quantum number, **s**, which describes the spin of each electron (spin up or spin down). The number **s** can be +1/2 or −1/2.

The Pauli exclusion principle states that no two electrons can occupy the same quantum state: every electron in an atom must have a unique combination of quantum numbers.

The above conventions imply a preferred axis (for example, the *z* direction in Cartesian coordinates), and they also imply a preferred direction along this preferred axis. Otherwise there would be no sense in distinguishing $m = +1$ from $m = -1$. As such, the model is most useful when applied to physical systems that share these symmetries. The Stern–Gerlach experiment — where an atom is exposed to a magnetic field — provides one such example.

Real Orbitals

An atom that is embedded in a crystalline solid feels multiple preferred axes, but no preferred direction. Instead of building atomic orbitals out of the product of radial functions and a single spherical harmonic, linear combinations of spherical harmonics are typically used, designed so that the imaginary part of the spherical harmonics cancel out. These **real orbitals** are the building blocks most commonly shown in orbital visualizations.

In the real hydrogen-like orbitals, for example, *n* and *l* have the same interpretation and significance as their complex counterparts, but *m* is no longer a good quantum number (though its absolute value is). The orbitals are given new names based on their shape with respect to a standardized Cartesian basis. The real hydrogen-like p orbitals are given by the following

$$p_z = p_0$$

$$p_x = \frac{1}{\sqrt{2}}(p_1 + p_{-1})$$

$$p_y = \frac{1}{i\sqrt{2}}(p_1 - p_{-1})$$

where $p_0 = R_{n1} Y_{10}$, $p_1 = R_{n1} Y_{11}$, and $p_{-1} = R_{n1} Y_{1-1}$, are the complex orbitals corresponding to $l = 1$.

The equations for the p_x and p_y orbitals depend on the phase convention used for the spherical harmonics. The above equations suppose that the spherical harmonics are defined by $Y_\ell^m(\theta,\varphi) = Ne^{im\varphi}P_\ell^m(\cos\theta)$. However some quantum physicists include a phase factor $(-1)^m$in these definitions, which has the effect of relating the p_x orbital to a *difference* of spherical harmonics and the p_y orbital to the corresponding *sum*. (For more detail, see Spherical harmonics#Conventions).

SHAPES OF ORBITALS

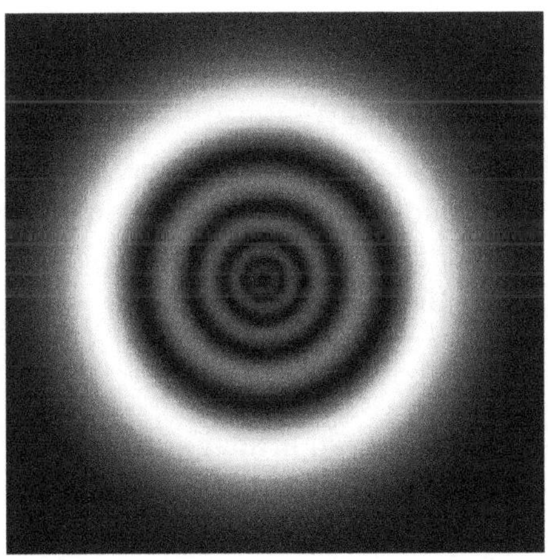

Cross-section of computed hydrogen atom orbital $(\psi(r,\theta,\varphi)^2)$ for the 6s ($n = 6$, $l = 0$, $m = 0$) orbital. Note that s orbitals, though spherically symmetrical, have radially placed wave-nodes for $n > 1$. However, only s orbitals invariably have a center anti-node; the other types never do.

Simple pictures showing orbital shapes are intended to describe the angular forms of regions in space where the electrons occupying the orbital are likely to be found. The diagrams cannot, however, show the entire region where an electron can be found, since according to quantum mechanics there is a non-zero probability of finding the electron (almost) anywhere in space. Instead the diagrams are approximate representations of boundary or contour surfaces where the probability density $|\psi(r,\theta,\varphi)|^2$ has a constant value, chosen so that there is a certain probability (for example 90%) of finding the electron within the contour. Although $|\psi|^2$ as the square of an absolute value is everywhere non-negative, the sign of the wave function $\psi(r,\theta,\varphi)$ is often indicated in each subregion of the orbital picture.

Sometimes the ψ function will be graphed to show its phases, rather than the $|\psi(r,\theta,\varphi)|^2$ which shows probability density but has no phases (which have been lost in the process of taking the absolute value, since $\psi(r,\theta,\varphi)$ is a complex number). $|\psi(r,\theta,\varphi)|^2$ orbital graphs tend to have less spherical, thinner lobes than $\psi(r,\theta,\varphi)$ graphs, but have the same number of lobes in the same places, and otherwise are recognizable. This article, in order to show wave function phases, shows mostly $\psi(r,\theta,\varphi)$ graphs.

The lobes can be viewed as standing wave interference patterns between the two counter rotating, ring resonant travelling wave "*m*" and "−*m*" modes, with the projection of the orbital onto the xy plane having a resonant "*m*" wavelengths around the circumference. Though rarely depicted the travelling wave solutions can be viewed as rotating banded tori, with the bands representing phase information. For each m there are two standing wave solutions $\langle m\rangle+\langle -m\rangle$ and $\langle m\rangle-\langle -m\rangle$. For the case where $m = 0$ the orbital is vertical, counter rotating information is unknown, and the orbital is z-axis symmetric. For the case where $l = 0$ there are no counter rotating modes. There are only radial modes and the shape is spherically symmetric. For any given n, the smaller l is, the more radial nodes there are. Loosely speaking n is energy, l is analogous to eccentricity, and m is orientation. For the record, in the classical case, a ring resonant travelling wave, for example in a circular transmission line, unless actively forced, will spontaneously decay into a ring resonant standing wave because reflections will build up over time at even the smallest imperfection or discontinuity.

Generally speaking, the number n determines the size and energy of the orbital for a given nucleus: as $n$increases, the size of the orbital increases. However, in comparing different elements, the higher nuclear charge Z of heavier elements causes their orbitals to contract by comparison to lighter ones, so that the overall size of the whole atom remains very roughly constant, even as the number of electrons in heavier elements (higher Z) increases.

Also in general terms, l determines an orbital's shape, and $m_l$its orientation. However, since some orbitals are described by equations in complex numbers, the shape sometimes depends on m_l also. Together, the whole set of orbitals for a given l and n fill space as symmetrically as possible, though with increasingly complex sets of lobes and nodes.

The single s-orbitals $\ell = 0$ are shaped like spheres. For $n = 1$it is roughly a solid ball (it is most dense at the center and fades exponentially outwardly), but for $n = 2$ or more, each single s-orbital is composed of spherically symmetric surfaces which are nested shells (i.e., the "wave-structure" is radial, following a sinusoidal radial component as well). See illustration of a cross-section of these nested shells, at right. The s-orbitals for all $n$numbers are the only orbitals with an anti-node (a region of high wave function density) at the center of the nucleus. All other orbitals (p, d, f, etc.) have angular momentum, and thus avoid the nucleus (having a wave node *at* the nucleus). Recently, there has been an effort to experimentally image the 1*s* and 2*p* orbitials in a $SrTiO_3$ crystal using scanning transmission electron microscopy with energy dispersive x-ray spectroscopy.Because the imaging was conducted using an electron beam, Coulombic beam-orbital interaction that is often termed as the impact parameter effect is included in the final outcome (see the figure at right).

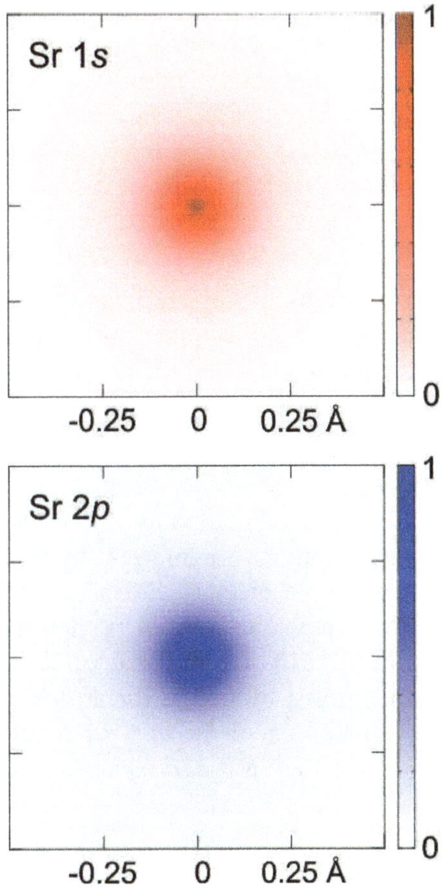

Experimentally imaged 1*s* and 2*p* core-electron orbitals of Sr, including the effects of atomic thermal vibrations and excitation broadening, retrieved from energy dispersive x-ray spectroscopy (EDX) in scanning transmission electron microscopy (STEM).

The shapes of p, d and f-orbitals are described verbally here and shown graphically in the *Orbitals table* below. The three p-orbitals for $n = 2$ have the form of two ellipsoids with a point of tangency at the nucleus (the two-lobed shape is sometimes referred to as a "dumbbell"—there are two lobes pointing in opposite directions from each other). The three p-orbitals in each shell are oriented at right angles to each other, as determined by their respective linear combination of values of m_l. The overall result is a lobe pointing along each direction of the primary axes.

Four of the five d-orbitals for $n = 3$ look similar, each with four pear-shaped lobes, each lobe tangent at right angles to two others, and the centers of all four lying in one plane. Three of these planes are the xy-, xz-, and yz-planes—the lobes are between the pairs of primary axes—and the fourth has the centres along the x and y axes themselves. The fifth and final d-orbital consists of three regions of high probability density: a torus with two pear-shaped regions placed symmetrically on its z axis. The overall total of 18 directional lobes point in every primary axis direction and between every pair.

There are seven f-orbitals, each with shapes more complex than those of the d-orbitals.

Additionally, as is the case with the s orbitals, individual p, d, f and g orbitals with *n* values higher than the lowest possible value, exhibit an additional radial node structure which is reminiscent of harmonic waves of the same type, as compared with the lowest (or fundamental) mode of the wave. As with s orbitals, this phenomenon provides p, d, f, and g orbitals at the next higher possible value of *n* (for example, 3p orbitals vs. the fundamental 2p), an additional node in each lobe. Still higher values of *n*further increase the number of radial nodes, for each type of orbital.

The shapes of atomic orbitals in one-electron atom are related to 3-dimensional spherical harmonics. These shapes are not unique, and any linear combination is valid, like a transformation to cubic harmonics, in fact it is possible to generate sets where all the d's are the same shape, just like the p_x, p_y,and p_z are the same shape.

Orbitals Table

This table shows all orbital configurations for the real hydrogen-like wave functions up to 7s, and therefore covers the simple electronic configuration for all elements in the periodic table up to radium. "ψ" graphs are shown with − and + wave function phases shown in two different colors (arbitrarily red and blue). The p_z orbital is the same as the p_0 orbital, but the p_x and p_y are formed by taking linear combinations of the p_{+1} and p_{-1} orbitals (which is why they are listed under the $m = \pm 1$ label). Also, the p_{+1} and p_{-1} are not the same shape as the p_0, since they are pure spherical harmonics.

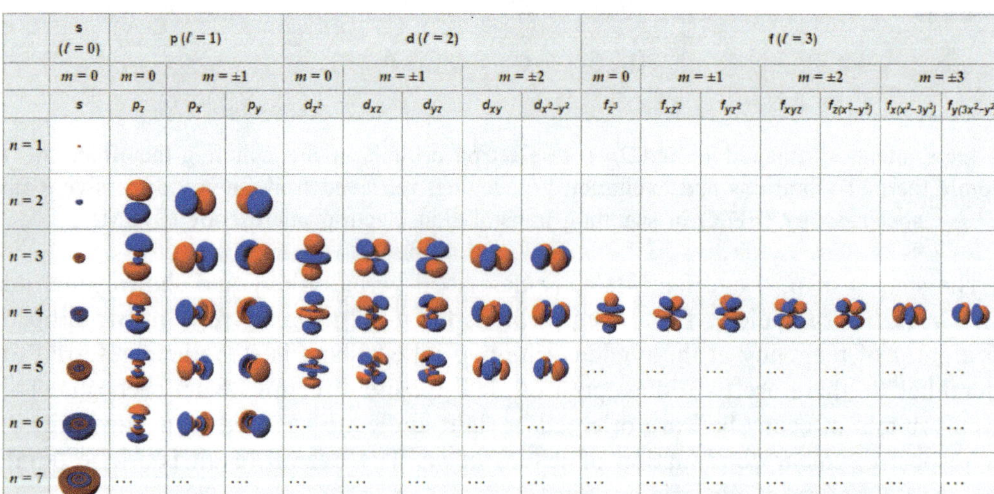

Qualitative Understanding of Shapes

The shapes of atomic orbitals can be understood qualitatively by considering the analogous case of standing waves on a circular drum. To see the analogy, the mean vibrational displacement of each bit of drum membrane from the equilibrium point over many cycles (a measure of average drum membrane velocity and momentum at that point) must be considered relative to that point's distance from the center of the drum head. If this displacement is taken as being analogous to the probability of finding an electron at a given distance from the nucleus, then it will be seen that the many modes of the vibrating disk form patterns that trace the various shapes of atomic orbitals. The basic reason for this correspondence lies in the fact that the distribution of kinetic energy and momentum in a matter-wave is predictive of where the particle associated with the wave will be. That is, the probability of finding an electron at a given place is also a function of the electron's average momentum at that point, since high electron momentum at a given position tends to "localize" the electron in that position, via the properties of electron wave-packets (see the Heisenberg uncertainty principle for details of the mechanism).

This relationship means that certain key features can be observed in both drum membrane modes and atomic orbitals. For example, in all of the modes analogous to **s** orbitals (the top row in the animated illustration below), it can be seen that the very center of the drum membrane vibrates most strongly, corresponding to the antinode in all **s** orbitals in an atom. This antinode means the electron is most likely to be at the physical position of the nucleus (which it passes straight through without scattering or striking it), since it is moving (on average) most rapidly at that point, giving it maximal momentum.

A mental "planetary orbit" picture closest to the behavior of electrons in **s** orbitals, all of which have no angular momentum, might perhaps be that of a Keplerian orbit with the orbital eccentricity of 1 but a finite major axis, not physically possible (because particles were to collide), but can be imagined as a limit of orbits with equal major axes but increasing eccentricity.

Below, a number of drum membrane vibration modes and the respective wave functions of the hydrogen atom are shown. A correspondence can be considered where the wave functions of a vibrating drum head are for a two-coordinate system $\psi(r, \theta)$ and the wave functions for a vibrating sphere are three-coordinate $\psi(r, \theta, \varphi)$.

s-type drum modes and wave functions

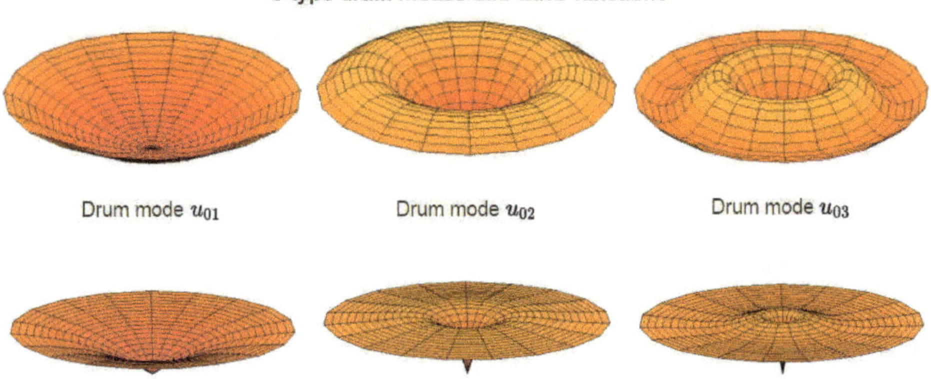

None of the other sets of modes in a drum membrane have a central antinode, and in all of them the center of the drum does not move. These correspond to a node at the nucleus for all non-**s** orbitals in an atom. These orbitals all have some angular momentum, and in the planetary model, they correspond to particles in orbit with eccentricity less than 1.0, so that they do not pass straight through the center of the primary body, but keep somewhat away from it.

In addition, the drum modes analogous to **p** and **d** modes in an atom show spatial irregularity along the different radial directions from the center of the drum, whereas all of the modes analogous to **s** modes are perfectly symmetrical in radial direction. The non radial-symmetry properties of non-**s** orbitals are necessary to localize a particle with angular momentum and a wave nature in an orbital where it must tend to stay away from the central attraction force, since any particle localized at the point of central attraction could have no angular momentum. For these modes, waves in the drum head tend to avoid the central point. Such features again emphasize that the shapes of atomic orbitals are a direct consequence of the wave nature of electrons.

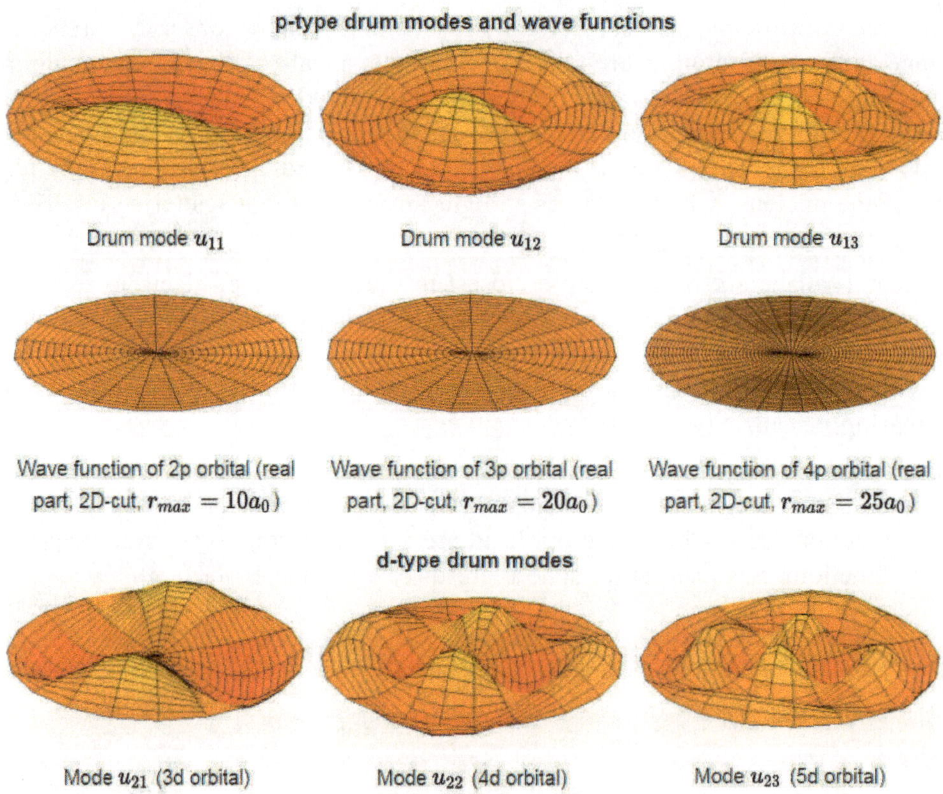

ORBITAL ENERGY

In atoms with a single electron (hydrogen-like atoms), the energy of an orbital (and, consequently, of any electrons in the orbital) is determined exclusively by n. The n

=1 orbital has the lowest possible energy in the atom. Each successively higher value of n, has a higher level of energy, but the difference decreases as n, increases. For high n, the level of energy becomes so high that the electron can easily escape from the atom. In single electron atoms, all levels with different ℓ within a given n are (to a good approximation) degenerate, and have the same energy. This approximation is broken to a slight extent by the effect of the magnetic field of the nucleus, and by quantum electrodynamics effects. The latter induce tiny binding energy differences especially for **s** electrons that go nearer the nucleus, since these feel a very slightly different nuclear charge, even in one-electron atoms; see Lamb shift.

In atoms with multiple electrons, the energy of an electron depends not only on the intrinsic properties of its orbital, but also on its interactions with the other electrons. These interactions depend on the detail of its spatial probability distribution, and so the energy levels of orbitals depend not only on n but also on ℓ. Higher values of ℓ are associated with higher values of energy; for instance, the 2p state is higher than the 2s state. When $\ell = 2$,, the increase in energy of the orbital becomes so large as to push the energy of orbital above the energy of the s-orbital in the next higher shell; when $\ell = 3$ the energy is pushed into the shell two steps higher. The filling of the 3d orbitals does not occur until the 4s orbitals have been filled.

The increase in energy for subshells of increasing angular momentum in larger atoms is due to electron–electron interaction effects, and it is specifically related to the ability of low angular momentum electrons to penetrate more effectively toward the nucleus, where they are subject to less screening from the charge of intervening electrons. Thus, in atoms of higher atomic number, the ℓ of electrons becomes more and more of a determining factor in their energy, and the principal quantum numbers n of electrons becomes less and less important in their energy placement.

The energy sequence of the first 35 subshells (e.g., 1s, 2p, 3d, etc.) is given in the following table. Each cell represents a subshell with n and ℓ given by its row and column indices, respectively. The number in the cell is the subshell's position in the sequence. For a linear listing of the subshells in terms of increasing energies in multielectron atoms, see the section below.

	s	p	d	f	g	h
1	1					
2	2	3				
3	4	5	7			
4	6	8	10	13		
5	9	11	14	17	*21*	
6	12	15	18	*22*	*26*	*31*
7	16	19	*23*	*27*	*32*	*37*
8	*20*	*24*	*28*	*33*	*38*	*44*
9	*25*	*29*	*34*	*39*	*45*	*51*
10	*30*	*35*	*40*	*46*	*52*	*59*

Note: empty cells indicate non-existent sublevels, while numbers in italics indicate sublevels that could (potentially) exist, but which do not hold electrons in any element currently known.

Electron Placement and the Periodic Table

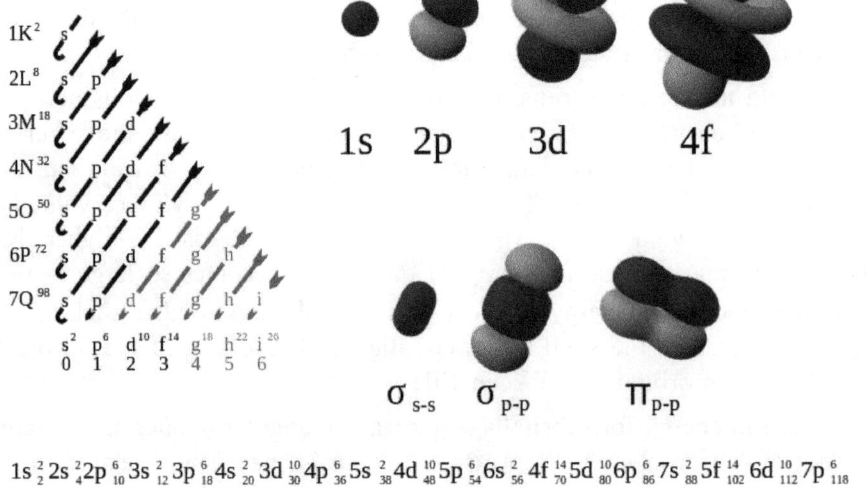

Electron atomic and molecular orbitals. The chart of orbitals (**left**) is arranged by increasing energy (see Madelung rule). *Note that atomic orbits are functions of three variables (two angles, and the distance r from the nucleus). These images are faithful to the angular component of the orbital, but not entirely representative of the orbital as a whole.*

Several rules govern the placement of electrons in orbitals (*electron configuration*). The first dictates that no two electrons in an atom may have the same set of values of quantum numbers (this is the Pauli exclusion principle). These quantum numbers include the three that define orbitals, as well as *s*, or spin quantum number. Thus, two electrons may occupy a single orbital, so long as they have different values of *s*. However, *only* two electrons, because of their spin, can be associated with each orbital.

Additionally, an electron always tends to fall to the lowest possible energy state. It is possible for it to occupy any orbital so long as it does not violate the Pauli exclusion principle, but if lower-energy orbitals are available, this condition is unstable. The electron will eventually lose energy (by releasing a photon) and drop into the lower orbital. Thus, electrons fill orbitals in the order specified by the energy sequence given above.

This behavior is responsible for the structure of the periodic table. The table may be divided into several rows (called ‹periods'), numbered starting with 1 at the top. The presently known elements occupy seven periods. If a certain period has number *i*, it consists of elements whose outermost electrons fall in the *i*th shell. Niels Bohr was the first to propose (1923) that the periodicity in the properties of the elements might be explained by the periodic filling of the electron energy levels, resulting in the electronic structure of the atom.

The periodic table may also be divided into several numbered rectangular ‹blocks'. The elements belonging to a given block have this common feature: their highest-energy electrons all belong to the same *l*-state (but the *n* associated with that *l*-state depends upon the period). For instance, the leftmost two columns constitute the 's-block'. The outermost electrons of Li and Be respectively belong to the 2s subshell, and those of Na and Mg to the 3s subshell.

The following is the order for filling the "subshell" orbitals, which also gives the order of the "blocks" in the periodic table:

1s, 2s, 2p, 3s, 3p, 4s, 3d, 4p, 5s, 4d, 5p, 6s, 4f, 5d, 6p, 7s, 5f, 6d, 7p

The "periodic" nature of the filling of orbitals, as well as emergence of the **s**, **p**, **d** and **f** "blocks", is more obvious if this order of filling is given in matrix form, with increasing principal quantum numbers starting the new rows ("periods") in the matrix. Then, each subshell (composed of the first two quantum numbers) is repeated as many times as required for each pair of electrons it may contain. The result is a compressed periodic table, with each entry representing two successive elements:

```
1s
2s                                                2p  2p  2p
3s                                                3p  3p  3p
4s                                3d  3d  3d  3d  3d  4p  4p  4p
5s                                4d  4d  4d  4d  4d  5p  5p  5p
6s  4f  4f  4f  4f  4f  4f  4f  5d  5d  5d  5d  5d  6p  6p  6p
7s  5f  5f  5f  5f  5f  5f  5f  6d  6d  6d  6d  6d  7p  7p  7p
```

Although this is the general order of orbital filling according to the Madelung rule, there are exceptions, and the actual electronic energies of each element are also dependent upon additional details of the atoms (see Electron configuration#Atoms: Aufbau principle and Madelung rule).

The number of electrons in an electrically neutral atom increases with the atomic number. The electrons in the outermost shell, or *valence electrons*, tend to be responsible for an element's chemical behavior. Elements that contain the same number of valence electrons can be grouped together and display similar chemical properties.

Relativistic Effects

For elements with high atomic number *Z*, the effects of relativity become more pronounced, and especially so for s electrons, which move at relativistic velocities as they penetrate the screening electrons near the core of high-*Z* atoms. This relativistic increase in momentum for high speed electrons causes a corresponding decrease in wavelength and contraction of 6s orbitals relative to 5d orbitals (by comparison to corresponding s and d electrons in lighter elements in the same column of the periodic table); this results in 6s valence electrons becoming lowered in energy.

Examples of significant physical outcomes of this effect include the lowered melting temperature of mercury (which results from 6s electrons not being available for metal bonding) and the golden color of gold and caesium (which results from narrowing of 6s to 5d transition energy to the point that visible light begins to be absorbed).

In the Bohr Model, an $n = 1$ electron has a velocity given by $v = Z\alpha c$, where Z is the atomic number, α is the fine-structure constant, and c is the speed of light. In non-relativistic quantum mechanics, therefore, any atom with an atomic number greater than 137 would require its 1s electrons to be traveling faster than the speed of light. Even in the Dirac equation, which accounts for relativistic effects, the wave function of the electron for atoms with $Z > 137$ is oscillatory and unbounded. The significance of element 137, also known as untriseptium, was first pointed out by the physicist Richard Feynman. Element 137 is sometimes informally called feynmanium (symbol Fy). However, Feynman's approximation fails to predict the exact critical value of Z due to the non-point-charge nature of the nucleus and very small orbital radius of inner electrons, resulting in a potential seen by inner electrons which is effectively less than Z. The critical Z value which makes the atom unstable with regard to high-field breakdown of the vacuum and production of electron-positron pairs, does not occur until Z is about 173. These conditions are not seen except transiently in collisions of very heavy nuclei such as lead or uranium in accelerators, where such electron-positron production from these effects has been claimed to be observed.

TRANSITIONS BETWEEN ORBITALS

Bound quantum states have discrete energy levels. When applied to atomic orbitals, this means that the energy differences between states are also discrete. A transition between these states (i.e., an electron absorbing or emitting a photon) can thus only happen if the photon has an energy corresponding with the exact energy difference between said states.

Consider two states of the hydrogen atom:

State 1) $n = 1$, $l = 0$, $m_l = 0$ and $s = +1/2$

State 2) $n = 2$, $l = 0$, $m_l = 0$ and $s = +1/2$

By quantum theory, state 1 has a fixed energy of E_1, and state 2 has a fixed energy of E_2. Now, what would happen if an electron in state 1 were to move to state 2? For this to happen, the electron would need to gain an energy of exactly $E_2 - E_1$. If the electron receives energy that is less than or greater than this value, it cannot jump from state 1 to state 2. Now, suppose we irradiate the atom with a broad-spectrum of light. Photons that reach the atom that have an energy of exactly $E_2 - E_1$ will be absorbed by the electron in state 1, and that electron will jump to state 2. However, photons that are greater or lower in energy cannot be absorbed by the electron, because the electron can only jump to one of the orbitals, it cannot jump to a state between orbitals. The result is that only photons of a specific frequency will be absorbed by the atom. This creates a line in the spectrum, known as an absorption line, which corresponds to the energy difference between states 1 and 2.

The atomic orbital model thus predicts line spectra, which are observed experimentally. This is one of the main validations of the atomic orbital model.

The atomic orbital model is nevertheless an approximation to the full quantum theory, which only recognizes many electron states. The predictions of line spectra are qualitatively useful but are not quantitatively accurate for atoms and ions other than those containing only one electron.

ATOMS IN MOLECULES

The quantum theory of atoms in molecules (QTAIM) is a model of molecular and condensed matter electronic systems (such as crystals) in which the principal objects of molecular structure - atoms and bonds - are natural expressions of a system's observable electron density distribution function. An electron density distribution of a molecule is a probability distribution that describes the average manner in which the electronic charge is distributed throughout real space in the attractive field exerted by the nuclei. According to QTAIM, molecular structure is revealed by the stationary points of the electron density together with the gradient paths of the electron density that originate and terminate at these points. QTAIM was primarily developed by Professor Richard Bader and his research group at McMaster University over the course of decades, beginning with analyses of theoretically calculated electron densities of simple molecules in the early 1960s and culminating with analyses of both theoretically and experimentally measured electron densities of crystals in the 90s. The development of QTAIM was driven by the assumption that, since the concepts of atoms and bonds have been and continue to be so ubiquitously useful in interpreting, classifying, predicting and communicating chemistry, they should have a well-defined physical basis.

QTAIM recovers the central operational concepts of the molecular structure hypothesis, that of a functional grouping of atoms with an additive and characteristic set of properties, together with a definition of the bonds that link the atoms and impart the structure. QTAIM defines chemical bonding and structure of a chemical system based on the topology of the electron density. In addition to bonding, AIM allows the calculation of certain physical properties on a per-atom basis, by dividing space up into atomic volumes containing exactly one nucleus, which acts as a local attractor of the electron density. In QTAIM an atom is defined as a *proper open system*, i.e. a system that can share energy and electron density which is localized in the 3D space. The mathematical study of these features is usually referred to in the literature as charge density *topology*.

QTAIM rests on the fact that the dominant topological property of the vast majority of electron density distributions is the presence of strong maxima that occur exclusively at the nuclei, certain pairs of which are linked together by ridges of electron density. In terms of an electron density distribution's gradient vector field, this corresponds to a complete, non-overlapping partitioning of a molecule into three-dimensional basins (atoms) that are linked together by shared two-dimensional separatrices (interatomic surfaces). Within each interatomic surface, the electron density is a maximum at the corresponding internuclear saddle point, which also lies at the minimum of the ridge between corresponding pair of nuclei, the ridge being defined by the pair of gradient trajectories (bond path) originating at the saddle point and terminating at the nuclei. Because QTAIM atoms are always bounded by surfaces having zero flux in the gradient vector field of the electron density, they have some unique quantum mechanical properties compared to other subsystem definitions, including unique electronic kinetic energy, the satisfaction of an electronic virial theorem analogous to the molecular electronic virial theorem and some interesting variational properties.

QTAIM has gradually become a method for addressing possible questions regarding chemical systems, in a variety of situations hardly handled before by any other model or theory in Chemistry

Applications

QTAIM is applied to the description of certain organic crystals with unusually short distances between neighboring molecules as observed by X-ray diffraction. For example, in the crystal structure of molecular chlorine the experimental Cl...Cl distance between two molecules is 327 picometres which is less than the sum of the van der Waals radii of 350 picometres. In one QTAIM result 12 bond paths start from each chlorine atom to other chlorine atoms including the other chlorine atom in the molecule. The theory also aims to explain the metallic properties of metallic hydrogen in much the same way.

The theory is also applied to so-called **hydrogen-hydrogen bonds** as they occur in molecules such as phenanthrene and chrysene. In these compounds the distance between two ortho hydrogen atoms again is shorter than their van der Waals radii and according to in silico experiments based on this theory, a bond path is identified between them. Both hydrogen atoms have identical electron density and are closed shell and therefore they are very different from the so-called dihydrogen bonds which are postulated for compounds such as $(CH_3)_2NHBH_3$ and also different from so-called agostic interactions.

In mainstream chemistry close proximity of two nonbonding atoms leads to destabilizing steric repulsion but in QTAIM the observed hydrogen hydrogen interactions are in fact stabilizing. It is well known that both kinked phenanthrene and chrysene are around 6 kcal/mol (25 kJ/mol) more stable than their linear isomersanthracene and tetracene. One traditional explanation is given by Clar's rule. QTAIM shows that a calculated stabilization for phenanthrene by 8 kcal/mol (33 kJ/mol) is the result of destabilization of the compound by 8 kcal/mol (33 kJ/mol) originating from electron transfer from carbon to hydrogen, offset by 12.1 kcal (51 kJ/mol) of stabilization due to a H..H bond path. The electron density at the critical point between the two hydrogen atoms is low, 0.012 e for phenanthrene. Another property of the bond path is its curvature.

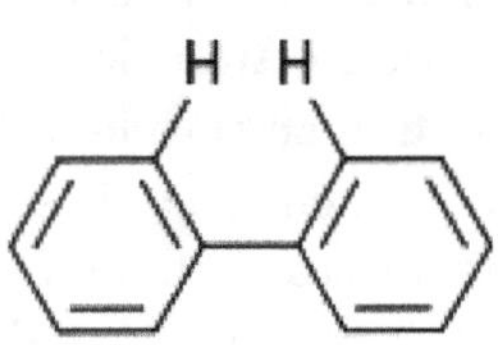

1

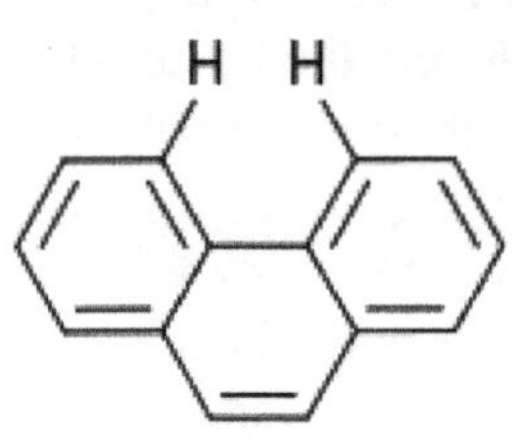

2

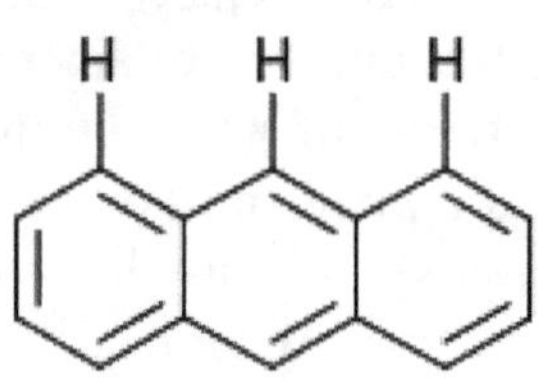

3

Another molecule studied in QTAIM is biphenyl. Its two phenyl rings are oriented at a 38° angle with respect to each other with the planar molecular geometry (encountered in a rotation around the central C-C bond) destabilized by 2.1 kcal/mol (8.8 kJ/mol) and the perpendicular one destabilized by 2.5 kcal/mol (10.5 kJ/mol). The classic explanations for this rotation barrier are steric repulsion between the ortho-hydrogen atoms (planar) and breaking of delocalization of pi density over both rings (perpendicular).

In QTAIM the energy increase on decreasing the dihedral angle from 38° to 0° is a summation of several factors. Destabilizing factors are the increase in bond length between the connecting carbon atoms (because they have to accommodate the approaching hydrogen atoms) and transfer of electronic charge from carbon to hydrogen. Stabilizing factors are increased delocalization of pi-electrons from one ring to the other and the one that tips the balance is a hydrogen - hydrogen bond between the ortho hydrogens.

The hydrogen bond is not without its critics. According to one the relative stability of phenanthrene compared to its isomers can be adequately explained by comparing resonance stabilizations. Another critic argues that the stability of phenanthrene can be attributed to more effective pi-pi overlap in the central double bond; the existence of bond paths are not questioned but the stabilizing energy derived from it is.

REFERENCES

1. Orchin, Milton; Macomber, Roger S.; Pinhas, Allan; Wilson, R. Marshall (2005). Atomic Orbital Theory (PDF).
2. Daintith, J. (2004). Oxford Dictionary of Chemistry. New York: Oxford University Press. ISBN 0-19-860918-3.
3. Griffiths, David (1995). Introduction to Quantum Mechanics. Prentice Hall. pp. 190–191. ISBN 0-13-124405-1.
4. Levine, Ira (2000). Quantum Chemistry (5 ed.). Prentice Hall. pp. 144–145. ISBN 0-13-685512-1.
5. Laidler, Keith J.; Meiser, John H. (1982). Physical Chemistry. Benjamin/ Cummings. p. 488. ISBN 0-8053-5682-7.
6. Atkins, Peter; de Paula, Julio; Friedman, Ronald (2009). Quanta, Matter, and Change: A Molecular Approach to Physical Chemistry. Oxford University Press. p. 106. ISBN 978-0-19-920606-3.
7. Feynman, Richard; Leighton, Robert B.; Sands, Matthew (2006). The Feynman Lectures on Physics -The Definitive Edition, Vol 1 lect 6. Pearson PLC, Addison Wesley. p. 11. ISBN 0-8053-9046-4.
8. Roger Penrose, The Road to Reality
9. Mulliken, Robert S. (July 1932). "Electronic Structures of Polyatomic Molecules and Valence. II. General Considerations". Physical Review. **41** *(1): 49–71. Bibcode:1932PhRv...41...49M. doi:10.1103/PhysRev.41.49.*
10. Bohr, Niels (1913). "On the Constitution of Atoms and Molecules". Philosophical Magazine. **26** *(1): 476. Bibcode:1914Natur..93..268N. doi:10.1038/093268a0.*

11. Nagaoka, Hantaro (May 1904). "Kinetics of a System of Particles illustrating the Line and the Band Spectrum and the Phenomena of Radioactivity". Philosophical Magazine. **7** *(41): 445–455. doi:10.1080/14786440409463141.*

12. Bryson, Bill (2003). A Short History of Nearly Everything. Broadway Books. pp. 141–143. ISBN 0-7679-0818-X.

13. Thomson, J. J. (1897). "Cathode rays". Philosophical Magazine. **44** *(269): 293. doi:10.1080/14786449708621070.*

14. Thomson, J. J. (1904). "On the Structure of the Atom: an Investigation of the Stability and Periods of Oscillation of a number of Corpuscles arranged at equal intervals around the Circumference of a Circle; with Application of the Results to the Theory of Atomic Structure" (extract of paper). Philosophical Magazine Series 6. **7** *(39): 237. doi:10.1080/14786440409463107.*

15. Rhodes, Richard (1995). The Making of the Atomic Bomb. Simon & Schuster. pp. 50–51. ISBN 978-0-684-81378-3.

16. Nagaoka, Hantaro (May 1904). "Kinetics of a System of Particles illustrating the Line and the Band Spectrum and the Phenomena of Radioactivity". Philosophical Magazine. **7** *(41): 446. doi:10.1080/14786440409463141.*

17. Heisenberg, W. (March 1927). "Über den anschaulichen Inhalt der quantentheoretischen Kinematik und Mechanik". Zeitschrift für Physik A. **43** *(3–4): 172–198. Bibcode:1927ZPhy...43..172H. doi:10.1007/BF01397280.*

18. Bohr, Niels (April 1928). "The Quantum Postulate and the Recent Development of Atomic Theory". Nature. **121** *(3050): 580–590. Bibcode:1928Natur.121..580B. doi:10.1038/121580a0.*

19. Gerlach, W.; Stern, O. (1922). "Das magnetische Moment des Silberatoms". Zeitschrift für Physik. **9**: 353–355. Bibcode:1922ZPhy....9..353G. doi:10.1007/BF01326984.

20. Levine, Ira (2014). Quantum Chemistry (7th ed.). Pearson Education. pp. 141–2. ISBN 0-321-80345-0.

21. Blanco, Miguel A.; Flórez, M.; Bermejo, M. (December 1997). "Evaluation of the rotation matrices in the basis of real spherical harmonics". Journal of Molecular Structure: THEOCHEM. **419** *(1–3): 19–27. doi:10.1016/S0166-1280(97)00185-1.*

22. Messiah, Albert (1999). Quantum mechanics : two volumes bound as one (Two vol. bound as one, unabridged reprint ed.). Mineola, NY: Dover. ISBN 978-0-486-40924-5.

23. Claude Cohen-Tannoudji; Bernard Diu; Franck Laloë; et al. (1996). Quantum mechanics. Translated by from the French by Susan Reid Hemley. Wiley-Interscience. ISBN 978-0-471-56952-7.

24. Jeong, Jong Seok; Odlyzko, Michael L.; Xu, Peng; Jalan, Bharat; Mkhoyan, K. Andre (2016-04-26). "Probing core-electron orbitals by scanning transmission electron microscopy and measuring the delocalization of core-level excitations". Physical Review B. **93** *(16): 165140. doi:10.1103/PhysRevB.93.165140.*

25. Powell, Richard E. (1968). "The five equivalent d orbitals". Journal of Chemical Education. **45** *(1): 45. Bibcode:1968JChEd..45...45P. doi:10.1021/ed045p45.*

26. Kimball, George E. (1940). "Directed Valence". The Journal of Chemical Physics. **8** *(2): 188. Bibcode:1940JChPh...8..188K. doi:10.1063/1.1750628.*

27. Cazenave, Lions, T., P.; Lions, P. L. (1982). "Orbital stability of standing waves for some nonlinear Schrödinger equations". Communications in Mathematical Physics. **85** *(4): 549–561. Bibcode:1982CMaPh..85..549C. doi:10.1007/BF01403504.*

28. Bohr, Niels (1923). "Über die Anwendung der Quantumtheorie auf den Atombau. I". Zeitschrift für Physik. **13**: 117. Bibcode:1923ZPhy...13..117B. doi:10.1007/BF01328209.

29. Lower, Stephen. "Primer on Quantum Theory of the Atom".

30. Szabo, Attila (1969). "Contour diagrams for relativistic orbitals". Journal of Chemical Education. **46** *(10): 678. Bibcode:1969JChEd..46..678S. doi:10.1021/ed046p678.*

CHAPTER 5

Molecular Orbital Theory

Chapter Contents

- **Introduction**
- **Quantitative Applications**
- **History**
- **Types of Orbitals**
- **Overview**
- **Molecular Orbital**
- **Molecular Orbital Diagram**
- **Molecular Orbital Examples**

Modern chemistry has depended upon the use of models of increasing comlexity. Atoms can be represented as spheres connected by cyclinders or sticks. In order to understand the mechanism of many reactions, Lewis Theory, developed by Robinson and Ingold, can provide a succesful answer.

Lewis Theory uses curly arrows to denote electron migration during a chemical reaction and has led to a greater understanding of the factors controlling chemical reactions.

Pauling with others, developed Resonance Theory, which provided the rationale to an all-embracing orbital theory. The use of "canonical forms" and "resonance hybrids", alonng with extensive use of curvy arrows has provided the fundamental background to modern organic theory, but for eg. Diels-Alder and pericyclic reactions, the curly arrow format is not very clear and in some instances the reactions are described as no-machanism reactions. Woodward and Hoffmann showed that by examining the interaction of the frontier molecular orbitals (ie. the Highest Occupied, HOMO and Lowest Unoccupied, LUMO) both the regio- and stereospecificity could be accountred for. Woodward and Hoffmann work was assimilated into general organic reaction theory.

INTRODUCTION

In chemistry, molecular orbital (MO) theory is a method for determining molecular structure in which electrons are not assigned to individual bonds between atoms, but are treated as moving under the influence of the nuclei in the whole molecule. The spatial and energetic properties of electrons within atoms are fixed by quantum mechanics to form orbitals that contain these electrons. While atomic orbitals contain electrons ascribed to a single atom, molecular orbitals, which surround a number of atoms in a molecule, contain valence electrons between atoms. Molecular orbital theory, which was proposed in the early twentieth century, revolutionized the study of bonding by approximating the positions of bonded electrons—the molecular orbitals—as linear combinations of atomic orbitals (LCAO). These approximations are now made by applying the density functional theory (DFT) or Hartree–Fock (HF) models to the Schrödinger equation.

QUANTITATIVE APPLICATIONS

In this theory, each molecule has a set of molecular orbitals, in which it is assumed that the molecular orbital wave function ψ_j can be written as a simple weighted sum of the n constituent atomic orbitals χ_i, according to the following equation:

$$\psi_j = \sum_{i=1}^{n} c_{ij}\chi_i .$$

One may determine c_{ij} coefficients numerically by substituting this equation into the Schrödinger equation and applying the variational principle. The variational principle is a mathematical technique used in quantum mechanics to build up the coefficients of each atomic orbital basis. A larger coefficient means that the orbital basis is composed more of that particular contributing atomic orbital—hence, the molecular orbital is best characterized by that type. This method of quantifying orbital contribution as Linear Combinations of Atomic Orbitals is used in computational chemistry. An additional unitary transformation can be applied on the system to accelerate the convergence in some computational schemes. Molecular orbital theory was seen as a competitor to valence bond theory in the 1930s, before it was realized that the two methods are closely related and that when extended they become equivalent.

HISTORY

Molecular orbital theory was developed, in the years after valence bond theory had been established (1927), primarily through the efforts of Friedrich Hund, Robert Mulliken, John C. Slater, and John Lennard-Jones. MO theory was originally called the Hund-Mulliken theory. According to German physicist and physical chemist Erich Hückel, the first quantitative use of molecular orbital theory was the 1929 paper of Lennard-Jones. This paper notably predicted a triplet ground state for the dioxygen molecule which explained its paramagnetism (see Molecular orbital diagram#Dioxygen) before valence bond theory, which came up with its own explanation in 1931. The word orbital was introduced by

Mulliken in 1932. By 1933, the molecular orbital theory had been accepted as a valid and useful theory.

Erich Hückel applied molecular orbital theory to unsaturated hydrocarbon molecules starting in 1931 with his Hückel molecular orbital (HMO) method for the determination of MO energies for pi electrons, which he applied to conjugated and aromatic hydrocarbons. This method provided an explanation of the stability of molecules with six pi-electrons such as benzene.

The first accurate calculation of a molecular orbital wavefunction was that made by Charles Coulson in 1938 on the hydrogen molecule. By 1950, molecular orbitals were completely defined as eigenfunctions (wave functions) of the self-consistent field Hamiltonian and it was at this point that molecular orbital theory became fully rigorous and consistent. This rigorous approach is known as the Hartree–Fock method for molecules although it had its origins in calculations on atoms. In calculations on molecules, the molecular orbitals are expanded in terms of an atomic orbital basis set, leading to the Roothaan equations. This led to the development of many ab initio quantum chemistry methods. In parallel, molecular orbital theory was applied in a more approximate manner using some empirically derived parameters in methods now known as semi-empirical quantum chemistry methods.

The success of Molecular Orbital Theory also spawned ligand field theory, which was developed during the 1930s and 1940s as an alternative to crystal field theory.

TYPES OF ORBITALS

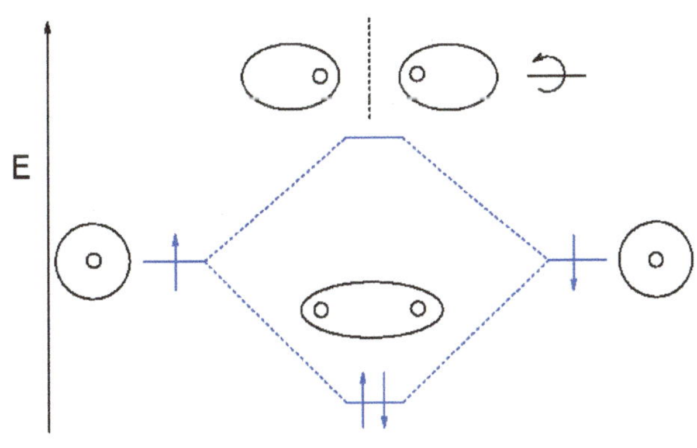

MO diagram showing the formation of molecular orbitals of H_2 (centre) from atomic orbitals of two H atoms. The lower-energy MO is bonding with electron density concentrated between the two H nuclei. The higher-energy MO is anti-bonding with electron density concentrated behind each H nucleus.

Molecular orbital (MO) theory uses a linear combination of atomic orbitals (LCAO) to represent molecular orbitals resulting from bonds between atoms. These are often divided into bonding orbitals, anti-bonding orbitals, and non-bonding orbitals. A bonding orbital concentrates electron density in the region between a given pair of atoms, so that its electron density will tend to attract each of the two nuclei toward the other and hold the two atoms together. An anti-bonding orbital concentrates electron density "behind" each nucleus (i.e. on the side of each atom which is farthest from the other atom), and so tends to pull each of the two nuclei away from the other and actually weaken the bond between the two nuclei. Electrons in non-bonding orbitals tend to be associated with atomic orbitals that do not interact positively or negatively with one another, and electrons in these orbitals neither contribute to nor detract from bond strength.

Molecular orbitals are further divided according to the types of atomic orbitals they are formed from. Chemical substances will form bonding interactions if their orbitals become lower in energy when they interact with each other. Different bonding orbitals are distinguished that differ by electron configuration (electron cloud shape) and by energy levels.

The molecular orbitals of a molecule can be illustrated in molecular orbital diagrams.

OVERVIEW

MO theory provides a global, delocalized perspective on chemical bonding. In MO theory, any electron in a molecule may be found anywhere in the molecule, since quantum conditions allow electrons to travel under the influence of an arbitrarily large number of nuclei, as long as they are in eigenstates permitted by certain quantum rules. Thus, when excited with the requisite amount of energy through high-frequency light or other means, electrons can transition to higher-energy molecular orbitals. For instance, in the simple case of a hydrogen diatomic molecule, promotion of a single electron from a bonding orbital to an antibonding orbital can occur under UV radiation. This promotion weakens the bond between the two hydrogen atoms and can lead to photodissociation—the breaking of a chemical bond due to the absorption of light.

Although in MO theory some molecular orbitals may hold electrons that are more localized between specific pairs of molecular atoms, other orbitals may hold electrons that are spread more uniformly over the molecule. Thus, overall, bonding is far more delocalized in MO theory, which makes it more applicable to resonant molecules that have equivalent non-integer bond orders than valence bond (VB) theory. This makes MO theory more useful for the description of extended systems.

An example is the MO description of benzene, C_6H_6, which is an aromatic hexagonal ring of six carbon atoms and three double bonds. In this molecule, 24 of the 30 total valence bonding electrons—24 coming from carbon atoms and 6 coming from hydrogen atoms—are located in 12 σ (sigma) bonding orbitals, which are located mostly between pairs of atoms (C-C or C-H), similarly to the electrons in the valence bond description. However, in benzene the remaining six bonding electrons are located in three π (pi) molecular bonding orbitals that are delocalized around the ring. Two of these electrons are in an MO that has equal orbital contributions from all six atoms. The other four electrons are in orbitals with vertical nodes at right angles to each other. As in the VB theory, all of these six delocalized π electrons reside in a larger space that exists above and below the ring plane. All carbon-carbon bonds in benzene are chemically equivalent. In MO theory this is a direct consequence of the fact that the three molecular π orbitals combine and evenly spread the extra six electrons over six carbon atoms.

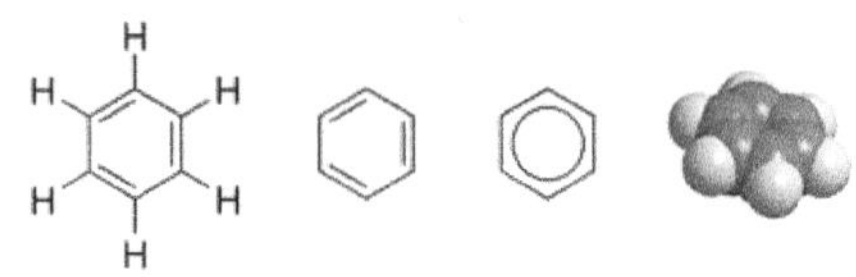

Structure of benzene.

In molecules such as methane, CH_4, the eight valence electrons are found in four MOs that are spread out over all five atoms. However, it is possible to transform the MOs into four localized sp^3 orbitals. Linus Pauling, in 1931, hybridized the carbon 2s and 2p orbitals so that they pointed directly at the hydrogen 1s basis functions and featured maximal overlap. However, the delocalized MO description is more appropriate for predicting ionization energies and the positions of spectral absorption bands. When methane is ionized, a single electron is taken from the valence MOs, which can come from the s bonding or the triply degenerate p bonding levels, yielding two ionization energies. In comparison, the explanation in VB theory is more complicated. When one electron is removed from an sp^3 orbital, resonance is invoked between four valence bond structures, each of which has a single one-electron bond and three two-electron bonds. Triply degenerate T_2 and A_1 ionized states (CH_4^+) are produced from different linear combinations of these four structures. The difference in energy between the ionized and ground state gives the two ionization energies.

As in benzene, in substances such as beta carotene, chlorophyll, or heme, some electrons in the π orbitals are spread out in molecular orbitals over long distances in a molecule, resulting in light absorption in lower energies (the visible spectrum), which accounts for the characteristic colours of these substances. This and other spectroscopic data for molecules are well explained in MO theory, with an emphasis on electronic states associated with multicenter orbitals, including mixing of orbitals premised on principles of orbital symmetry matching. The same MO principles also naturally explain some electrical phenomena, such as high electrical conductivity in the planar direction of the hexagonal atomic sheets that exist in graphite. This results from continuous band overlap of half-filled p orbitals and explains electrical conduction. MO theory recognizes that some electrons in the graphite atomic sheets are completely delocalized over arbitrary distances, and reside in very large molecular orbitals that cover an entire graphite sheet, and some electrons are thus as free to move and therefore conduct electricity in the sheet plane, as if they resided in a metal.

MOLECULAR ORBITAL

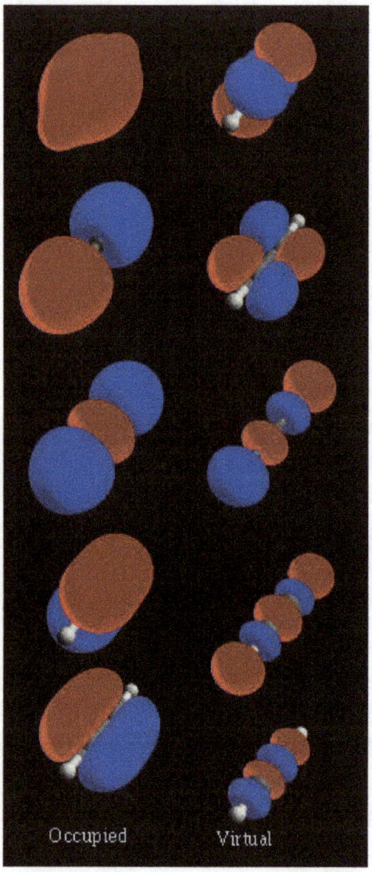

Complete acetylene (H–C≡C–H) molecular orbital set. The left column shows MO's which are occupied in the ground state, with the lowest-energy orbital at the top. The white and grey line visible in some MO's is the molecular axis passing through the nuclei. The orbital wave functions are positive in the red regions and negative in the blue. The right column shows virtual MO's which are empty in the ground state, but may be occupied in excited states.

In chemistry, a **molecular orbital (MO)** is a mathematical function describing the wave-like behavior of an electron in a molecule. This function can be used to calculate chemical and physical properties such as the probability of finding an electron in any specific region. The term orbital was introduced by Robert S. Mulliken in 1932 as an abbreviation for one-electron orbital wave function. At an elementary level, it is used to describe the region of space in which the function has a significant amplitude. Molecular orbitals are usually constructed by combining atomic orbitals or hybrid orbitals from each atom of the molecule, or other molecular orbitals from groups of atoms. They can be quantitatively calculated using the Hartree–Fock or self-consistent field (SCF) methods.

A molecular orbital (MO) can be used to represent the regions in a molecule where an electron occupying that orbital is likely to be found. Molecular orbitals are obtained from the combination of atomic orbitals, which predict the location of an electron in an atom. A molecular orbital can specify the electron configuration of a molecule: the spatial distribution and energy of one (or one pair of) electron(s). Most commonly a MO is represented as a linear combination of atomic orbitals (the LCAO-MO method), especially in qualitative or very approximate usage. They are invaluable in providing a simple model of bonding in molecules, understood through molecular orbital theory. Most present-day methods in computational chemistry begin by calculating the MOs of the system. A molecular orbital describes the behavior of one electron in the electric field generated by the nuclei and some average distribution of the other electrons. In the case of two electrons occupying the same orbital, the Pauli principle demands that they have opposite spin. Necessarily this is an approximation, and highly accurate descriptions of the molecular electronic wave function do not have orbitals (see configuration interaction).

Formation of Molecular Orbitals

Molecular orbitals arise from allowed interactions between atomic orbitals, which are allowed if the symmetries (determined from group theory) of the atomic orbitals are compatible with each other. Efficiency of atomic orbital interactions is determined from the overlap (a measure of how well two orbitals constructively interact with one another) between two atomic orbitals, which is significant if the atomic orbitals are close in energy. Finally, the number of molecular orbitals that form must equal the number of atomic orbitals in the atoms being combined to form the molecule.

Qualitative Discussion

For an imprecise, but qualitatively useful, discussion of the molecular structure, the molecular orbitals can be obtained from the "Linear combination of atomic orbitals molecular orbital method" ansatz. Here, the molecular orbitals are expressed as linear combinations of atomic orbitals.

Linear Combinations of Atomic Orbitals (LCAO)

Molecular orbitals were first introduced by Friedrich Hund and Robert S. Mulliken in 1927 and 1928. The linear combination of atomic orbitals or "LCAO" approximation for molecular orbitals was introduced in 1929 by Sir John Lennard-Jones. His ground-breaking paper showed how to derive the electronic structure of the fluorine and oxygen molecules from quantum principles. This qualitative approach to molecular orbital theory is part of the start of modern quantum chemistry. Linear combinations of atomic orbitals (LCAO) can be used to estimate the molecular orbitals that are formed upon bonding between the molecule›s constituent atoms. Similar to an atomic orbital, a Schrödinger equation, which describes the behavior of an electron, can be constructed for a molecular orbital as well. Linear combinations of atomic orbitals, or the sums and differences of the atomic wavefunctions, provide approximate solutions to the Hartree–Fock equations which correspond to the independent-particle approximation of the molecular Schrödinger equation. For simple diatomic molecules, the wavefunctions obtained are represented mathematically by the equations

$$\Psi = c_a\psi_a + c_b\psi_b$$
$$\Psi^* = c_a\psi_a - c_b\psi_b$$

where Ψ and Ψ^* are the molecular wavefunctions for the bonding and antibonding molecular orbitals, respectively, ψ_a and ψ_b are the atomic wavefunctions from atoms a and b, respectively, and c_a and c_b are adjustable coefficients. These coefficients can be positive or negative, depending on the energies and symmetries of the individual atomic orbitals. As the two atoms become closer together, their atomic orbitals overlap to produce areas of high electron density, and, as a consequence, molecular orbitals are formed between the two atoms. The atoms are held together by the electrostatic attraction between the positively charged nuclei and the negatively charged electrons occupying bonding molecular orbitals.

Bonding, Antibonding, and Nonbonding MOs

When atomic orbitals interact, the resulting molecular orbital can be of three types: bonding, antibonding, or nonbonding.

Bonding MOs:

- Bonding interactions between atomic orbitals are constructive (in-phase) interactions.
- Bonding MOs are lower in energy than the atomic orbitals that combine to produce them.

Antibonding MOs:

- Antibonding interactions between atomic orbitals are destructive (out-of-phase) interactions, with a nodal plane where the wavefunction of the antibonding orbital is zero between the two interacting atoms
- Antibonding MOs are higher in energy than the atomic orbitals that combine to produce them.

Nonbonding MOs:

- Nonbonding MOs are the result of no interaction between atomic orbitals because of lack of compatible symmetries.
- Nonbonding MOs will have the same energy as the atomic orbitals of one of the atoms in the molecule.

Sigma and pi Labels for MOs

The type of interaction between atomic orbitals can be further categorized by the molecular-orbital symmetry labels σ (sigma), π (pi), δ (delta), φ (phi), γ (gamma) etc. paralleling the symmetry of the atomic orbitals s, p, d, f and g. The number of nodal planes containing the internuclear axis between the atoms concerned is zero for σ MOs, one for π, two for δ, etc.

σ Symmetry

A MO with σ symmetry results from the interaction of either two atomic s-orbitals or two atomic p_z-orbitals. An MO will have σ-symmetry if the orbital is symmetric with respect to the axis joining the two nuclear centers, the internuclear axis. This means that rotation of the MO about the internuclear axis does not result in a phase change. A σ* orbital, sigma antibonding orbital, also maintains the same phase when rotated about the internuclear axis. The σ* orbital has a nodal plane that is between the nuclei and perpendicular to the internuclear axis.

π Symmetry

A MO with π symmetry results from the interaction of either two atomic p_x orbitals or p_y orbitals. An MO will have π symmetry if the orbital is asymmetric with respect to rotation about the internuclear axis. This means that rotation of the MO about the internuclear axis will result in a phase change. There is one nodal plane containing the internuclear axis, if real orbitals are considered.

A π* orbital, pi antibonding orbital, will also produce a phase change when rotated about the internuclear axis. The π* orbital also has a second nodal plane between the nuclei.

δ Symmetry

A MO with δ symmetry results from the interaction of two atomic d_{xy} or $d_{x^2-y^2}$ orbitals. Because these molecular orbitals involve low-energy d atomic orbitals, they are seen in transition-metal complexes. A δ bonding orbital has two nodal planes containing the internuclear axis, and a δ* antibonding orbital also has a third nodal plane between the nuclei.

φ Symmetry

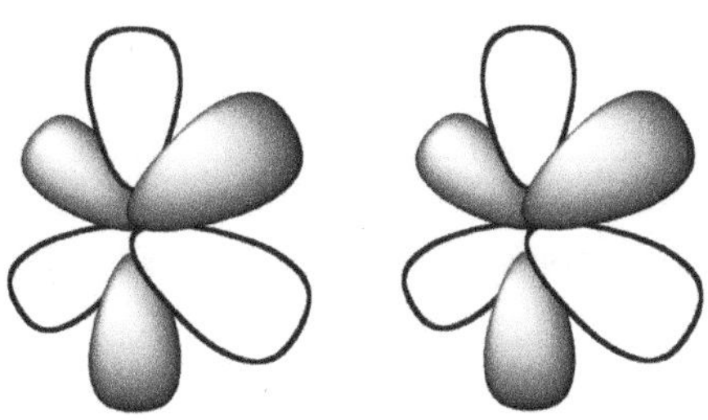

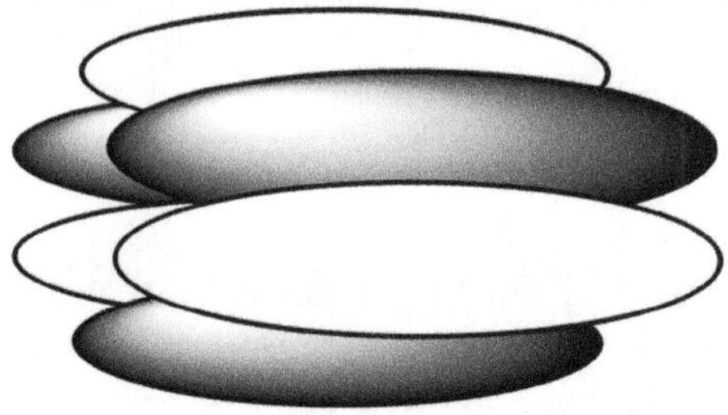

Suitably aligned f atomic orbitals overlap to form phi molecular orbital (a phi bond).

Theoretical chemists have conjectured that higher-order bonds, such as phi bonds corresponding to overlap of f atomic orbitals, are possible. There is as of 2005 only one known example of a molecule purported to contain a phi bond (a U−U bond, in the molecule U_2).

Gerade and Ungerade Symmetry

For molecules that possess a center of inversion (centrosymmetric molecules) there are additional labels of symmetry that can be applied to molecular orbitals. Centrosymmetric molecules include:

- Homonuclear diatomics, X_2
- Octahedral, EX_6
- Square planar, EX_4.

Non-centrosymmetric molecules include:

- Heteronuclear diatomics, XY
- Tetrahedral, EX_4.

If inversion through the center of symmetry in a molecule results in the same phases for the molecular orbital, then the MO is said to have gerade (g) symmetry, from the German word for even. If inversion through the center of symmetry in a molecule results in a phase change for the molecular orbital, then the MO is said to have ungerade (u) symmetry, from the German word for odd. For a bonding MO with σ-symmetry, the orbital is σ_g (s' + s'' is symmetric), while an antibonding MO with σ-symmetry the orbital is σ_u, because inversion of s' – s'' is antisymmetric. For a bonding MO with π-symmetry the orbital is π_u because inversion through the center of symmetry for would produce a sign change (the two p atomic orbitals are in phase with each other but the two lobes have opposite signs), while an antibonding MO with π-symmetry is π_g because inversion through the center of symmetry for would not produce a sign change (the two p orbitals are antisymmetric by phase).

MOLECULAR ORBITAL DIAGRAM

A **molecular orbital diagram**, or **MO diagram**, is a qualitative descriptive tool explaining chemical bonding in molecules in terms of molecular orbital theory in general and the linear combination of atomic orbitals (LCAO) molecular orbital method in particular. A fundamental principle of these theories is that as atoms bond to form molecules, a certain number of atomic orbitals combine to form the same number of molecular orbitals, although the electrons involved may be redistributed among the orbitals. This tool is very well suited for simple diatomic molecules such as dihydrogen, dioxygen, and carbon monoxide but becomes more complex when discussing even comparatively simple polyatomic molecules, such as methane. MO diagrams can explain why some molecules exist and others do not. They can also predict bond strength, as well as the electronic transitions that can take place.

Qualitative MO theory was introduced in 1928 by Robert S. Mulliken and Friedrich Hund. A mathematical description was provided by contributions from Douglas Hartree in 1928 and Vladimir Fock in 1930.

Basics

Molecular orbital diagrams are diagrams of molecular orbital (MO) energy levels, shown as short horizontal lines in the center, flanked by constituent atomic orbital (AO) energy levels for comparison, with the energy levels increasing from the bottom to the top. Lines, often dashed diagonal lines, connect MO levels with their constituent AO levels. Degenerate energy levels are commonly shown side by side. Appropriate AO and MO levels are filled with electrons by the Pauli Exclusion Principle, symbolized by small vertical arrows whose directions indicate the electron spins. The AO or MO shapes themselves are often not shown on these diagrams. For a diatomic molecule, an MO diagram effectively shows the energetics of the bond between the two atoms, whose AO unbonded energies are shown on the sides. For simple polyatomic molecules with a «central atom» such as methane (CH_4) or carbon dioxide (CO_2), a MO diagram may show one of the identical bonds to the central atom. For other polyatomic molecules, an MO diagram may show one or more bonds of interest in the molecules, leaving others out for simplicity. Often even for simple molecules, AO and MO levels of inner orbitals and their electrons may be omitted from a diagram for simplicity.

In MO theory molecular orbitals form by the overlap of atomic orbitals. Because σ bonds feature greater overlap than π bonds, σ and σ* bonding and antibonding orbitals feature greater energy splitting (separation) than π and π* orbitals. The atomic orbital energy correlates with electronegativity as more electronegative atoms hold their electrons more tightly, lowering their energies. MO modelling is only valid when the atomic orbitals have comparable energy; when the energies differ greatly the mode of bonding becomes ionic. A second condition for overlapping atomic orbitals is that they have the same symmetry.

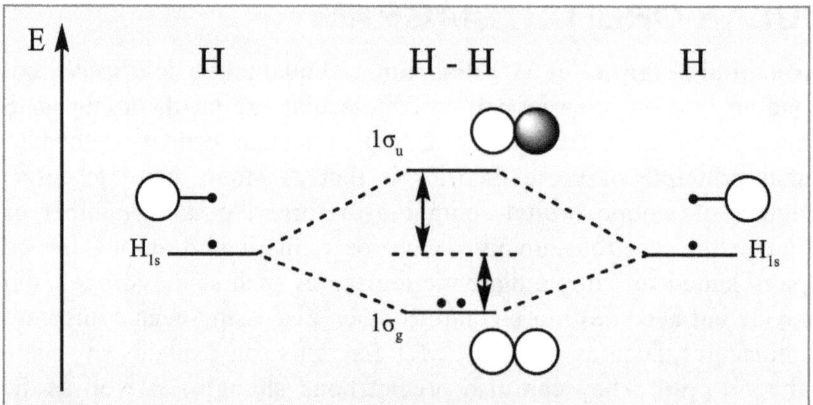

MO diagram for dihydrogen. Here electrons are shown by dots.

Two atomic orbitals can overlap in two ways depending on their phase relationship. The phase of an orbital is a direct consequence of the wave-like properties of electrons. In graphical representations of orbitals, orbital phase is depicted either by a plus or minus sign (which has no relationship to electric charge) or by shading one lobe. The sign of the phase itself does not have physical meaning except when mixing orbitals to form molecular orbitals.

Two same-sign orbitals have a constructive overlap forming a molecular orbital with the bulk of the electron density located between the two nuclei. This MO is called the bonding orbital and its energy is lower than that of the original atomic orbitals. A bond involving molecular orbitals which are symmetric with respect to rotation around the bond axis is called a sigma bond (**σ**-bond). If the phase changes, the bond becomes a pi bond (**π**-bond). Symmetry labels are further defined by whether the orbital maintains its original character after an inversion about its center; if it does, it is defined gerade, g. If the orbital does not maintain its original character, it is ungerade, u.

Atomic orbitals can also interact with each other out-of-phase which leads to destructive cancellation and no electron density between the two nuclei at the so-called nodal plane depicted as a perpendicular dashed line. In this anti-bonding MO with energy much higher than the original AO's, any electrons present are located in lobes pointing away from the central internuclear axis. For a corresponding **σ**-bonding orbital, such an orbital would be symmetrical but differentiated from it by an asterisk as in **σ***. For a **π**-bond, corresponding bonding and antibonding orbitals would not have such symmetry around the bond axis and be designated **π** and **π***, respectively.

The next step in constructing an MO diagram is filling the newly formed molecular orbitals with electrons. Three general rules apply:

- The Aufbau principle states that orbitals are filled starting with the lowest energy
- The Pauli exclusion principle states that the maximum number of electrons occupying an orbital is two, with opposite spins
- Hund›s rule states that when there are several MO's with equal energy, the electrons occupy the MO's one at a time before two electrons occupy the same MO.

The filled MO highest in energy is called the Highest Occupied Molecular Orbital or HOMO and the empty MO just above it is then the Lowest Unoccupied Molecular Orbital or LUMO. The electrons in the bonding MO's are called bonding electrons and any electrons in the antibonding orbital would be called antibonding electrons. The reduction in energy of these electrons is the driving force for chemical bond formation. Whenever mixing for an atomic orbital is not possible for reasons of symmetry or energy, a non-bonding MO is created, which is often quite similar to and has energy level equal or close to its constituent AO, thus not contributing to bonding energetics. The resulting electron configuration can be described in terms of bond type, parity and occupancy for example dihydrogen $1\sigma_g^2$. Alternatively it can be written as a molecular term symbol e.g. $^1\Sigma_g^+$ for dihydrogen. Sometimes, the letter **n** is used to designate a non-bonding orbital.

For a stable bond, the bond order, defined as

$$\text{Bond Order} = \frac{(\text{No. of electrons in bonding MOs}) - (\text{No. of electrons in anti-bonding MOs})}{2}$$

must be positive.

The relative order in MO energies and occupancy corresponds with electronic transitions found in photoelectron spectroscopy (PES). In this way it is possible to experimentally verify MO theory. In general, sharp PES transitions indicate nonbonding electrons and broad bands are indicative of bonding and antibonding delocalized electrons. Bands can resolve into fine structure with spacings corresponding to vibrational modes of the molecular cation (see Franck–Condon principle). PES energies are different from ionisation energies which relates to the energy required to strip off the nth electron after the first n − 1 electrons have been removed. MO diagrams with energy values can be obtained mathematically using the Hartree–Fock method. The starting point for any MO diagram is a predefined molecular geometry for the molecule in question. An exact relationship between geometry and orbital energies is given in Walsh diagrams.

s-p Mixing

The phenomenon of s-p mixing occurs when molecular orbitals of the same symmetry formed from the combination of 2s and 2p atomic orbitals are close enough in energy to further interact, which can lead to a change in the expected order of orbital energies. When molecular orbitals are formed, they are mathematically obtained from linear combinations of the starting atomic orbitals. Generally, in order to predict their relative energies, it is sufficient to consider only one atomic orbital from each atom to form a pair of molecular orbitals, as the contributions from the others are negligible. For instance, in dioxygen the $3\sigma_g$ MO can be roughly considered to be formed from interaction of oxygen $2p_z$ AOs only. It is found to be lower in energy than the $1\pi_u$ MO, both experimentally and from more sophisticated computational models, so that the expected order of filling is the $3\sigma_g$ before the $1\pi_u$. Hence the approximation to ignore the effects of further interactions is valid. However, experimental and computational results for homonuclear diatomics from Li_2 to N_2 and certain heteronuclear combinations such as CO and NO show that the $3\sigma_g$ MO is higher in energy than (and therefore filled after) the $1\pi_u$ MO. This can be rationalised as the first-approximation $3\sigma_g$ has a suitable symmetry to interact with the $2\sigma_g$ bonding MO formed from the 2s AOs. As a result, the $2\sigma_g$ is lowered in energy, whilst the $3\sigma_g$ is raised.

For the aforementioned molecules this results in the $3\sigma_g$ being higher in energy than the $1\pi_u$ MO, which is where s-p mixing is most evident. Likewise, interaction between the $2\sigma_u$* and $3\sigma_u$* MOs leads to a lowering in energy of the former and a raising in energy of the latter. However this is of less significance than the interaction of the bonding MOs.

Diatomic MO Diagrams

Dihydrogen

The smallest molecule, hydrogen gas exists as dihydrogen (H-H) with a single covalent bond between two hydrogen atoms. As each hydrogen atom has a single 1s atomic orbital for its electron, the bond forms by overlap of these two atomic orbitals. In figure 1 the two atomic orbitals are depicted on the left and on the right. The vertical axis always represents the orbital energies. Each atomic orbital is singly occupied with an up or down arrow representing an electron.

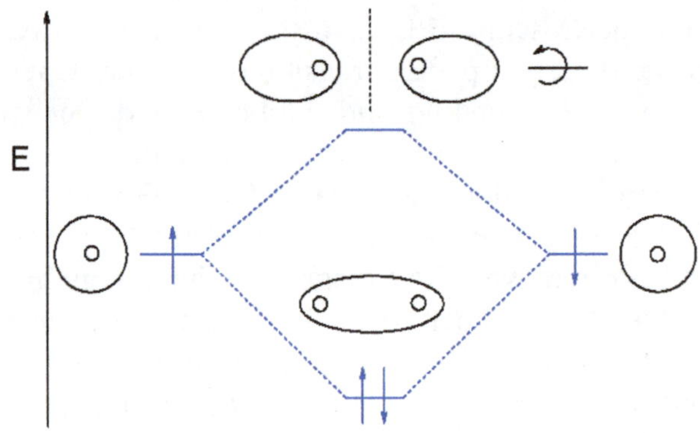

MO diagram of dihydrogen.

Application of MO theory for dihydrogen results in having both electrons in the bonding MO with electron configuration $1\sigma_g^2$. The bond order for dihydrogen is (2-0)/2 = 1. The photoelectron spectrum of dihydrogen shows a single set of multiplets between 16 and 18 eV (electron volts).

The dihydrogen MO diagram helps explain how a bond breaks. When applying energy to dihydrogen, a molecular electronic transition takes place when one electron in the bonding MO is promoted to the antibonding MO. The result is that there is no longer a net gain in energy.

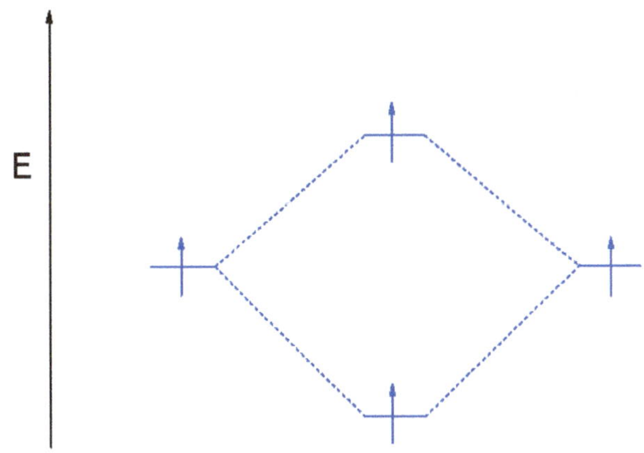

Bond breaking in MO diagram.

Dihelium and Diberyllium

Dihelium (He-He) is a hypothetical molecule and MO theory helps to explain why dihelium does not exist in nature. The MO diagram for dihelium looks very similar to that of dihydrogen, but each helium has two electrons in its 1s atomic orbital rather than one for hydrogen, so there are now four electrons to place in the newly formed molecular orbitals.

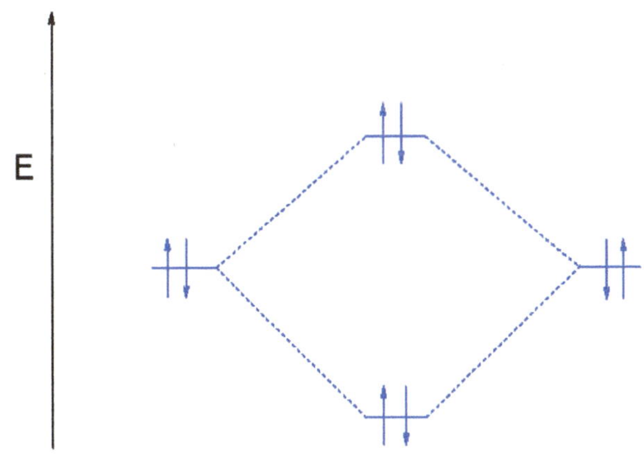

MO diagram of dihelium.

The only way to accomplish this is by occupying both the bonding and antibonding orbitals with two electrons, which reduces the bond order ((2−2)/2) to zero and cancels the net energy stabilization. However, by removing one electron from dihelium, the stable gas-phase species He_2^+ ion is formed with bond order 1/2.

Another molecule that is precluded based on this principle is **diberyllium**. Beryllium has an electron configuration $1s^2 2s^2$, so there are again two electrons in the valence level.

However, the 2s can mix with the 2p orbitals in diberyllium, whereas there are no p orbitals in the valence level of hydrogen or helium. This mixing makes the antibonding $1\sigma_u$ orbital slightly less antibonding than the bonding $1\sigma_g$ orbital is bonding, with a net effect that the whole configuration has a slight bonding nature. Hence the diberyllium molecule exists (and has been observed in the gas phase). It nevertheless still has a low dissociation energy of only 59 $kJ \cdot mol^{-1}$.

Dilithium

MO theory correctly predicts that dilithium is a stable molecule with bond order 1 (configuration $1\sigma_g^2 1\sigma_u^2 2\sigma_g^2$). The 1s MOs are completely filled and do not participate in bonding.

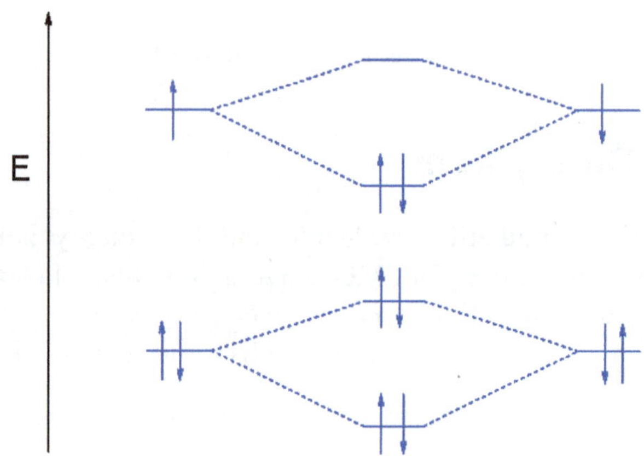

MO Diagram of Dilithium.

Dilithium is a gas-phase molecule with a much lower bond strength than dihydrogen because the 2s electrons are further removed from the nucleus. In a more detailed analysis both the 1σ orbitals have higher energies than the 1s AO and the occupied 2σ is also higher in energy than the 2s AO (see table 1).

Diboron

The MO diagram for diboron (B-B, electron configuration $1\sigma_g^2 1\sigma_u^2 2\sigma_g^2 2\sigma_u^2 1\pi_u^2$) requires the introduction of an atomic orbital overlap model for p orbitals. The three dumbbell-shaped p-orbitals have equal energy and are oriented mutually perpendicularly (or orthogonally). The p-orbitals oriented in the x-direction (p_x) can overlap end-on forming a bonding (symmetrical) σ orbital and an antibonding σ* molecular orbital. In contrast to the sigma 1s MO's, the σ 2p has some non-bonding electron density at either side of the nuclei and the σ* 2p has some electron density between the nuclei.

The other two p-orbitals, p_y and p_z, can overlap side-on. The resulting bonding orbital has its electron density in the shape of two lobes above and below the plane of the molecule. The orbital is not symmetric around the molecular axis and is therefore a pi orbital. The antibonding pi orbital (also asymmetrical) has four lobes pointing away from the nuclei. Both p_y and p_z orbitals form a pair of pi orbitals equal in energy (degenerate) and can have higher or lower energies than that of the sigma orbital.

In diboron the 1s and 2s electrons do not participate in bonding but the single electrons in the 2p orbitals occupy the $2\pi p_y$ and the $2\pi p_z$ MO's resulting in bond order 1. Because the electrons have equal energy (they are degenerate) diboron is a diradical and since the spins are parallel the compound is paramagnetic.

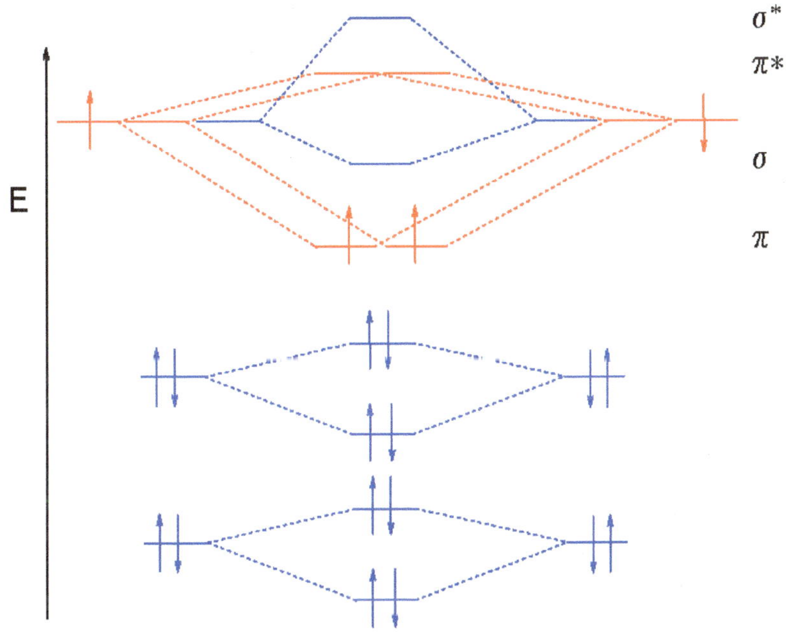

MO diagram of diboron.

In certain diborynes the boron atoms are excited and the bond order is 3.

icarbon

Like diboron, dicarbon (C-C electron configuration:$1\sigma_g^2 1\sigma_u^2 2\sigma_g^2 2\sigma_u^2 1\pi_u^4$) is a reactive gas-phase molecule. The molecule can be described as having two pi bonds but without a sigma bond.

Dinitrogen

The bond order for dinitrogen ($1\sigma_g^2 1\sigma_u^2 2\sigma_g^2 2\sigma_u^2 1\pi_u^4 3\sigma_g^2$) is three because two electrons are now also added in the 3σ MO. The MO diagram correlates with the experimental photoelectron spectrum for nitrogen. The 1σ electrons can be matched to a peak at 410 eV (broad), the $2\sigma_g$ electrons at 37 eV (broad), the $2\sigma_u$ electrons at 19 eV (doublet), the $1\pi_u^4$ electrons at 17 eV (multiplets), and finally the $3\sigma_g^2$ at 15.5 eV (sharp).

Dioxygen

MO treatment of dioxygen is different from that of the previous diatomic molecules because the pσ MO is now lower in energy than the 2π orbitals. This is attributed to interaction between the 2s MO and the $2p_z$MO. Distributing 8 electrons over 6 molecular orbitals leaves the final two electrons as a degenerate pair in the 2pπ* antibonding orbitals resulting in a bond order of 2. As in diboron, these two unpaired electrons have the same spin in the ground state, which is a paramagnetic diradical triplet oxygen. The first excited state has both HOMO electrons paired in one orbital with opposite spins, and is known as singlet oxygen.

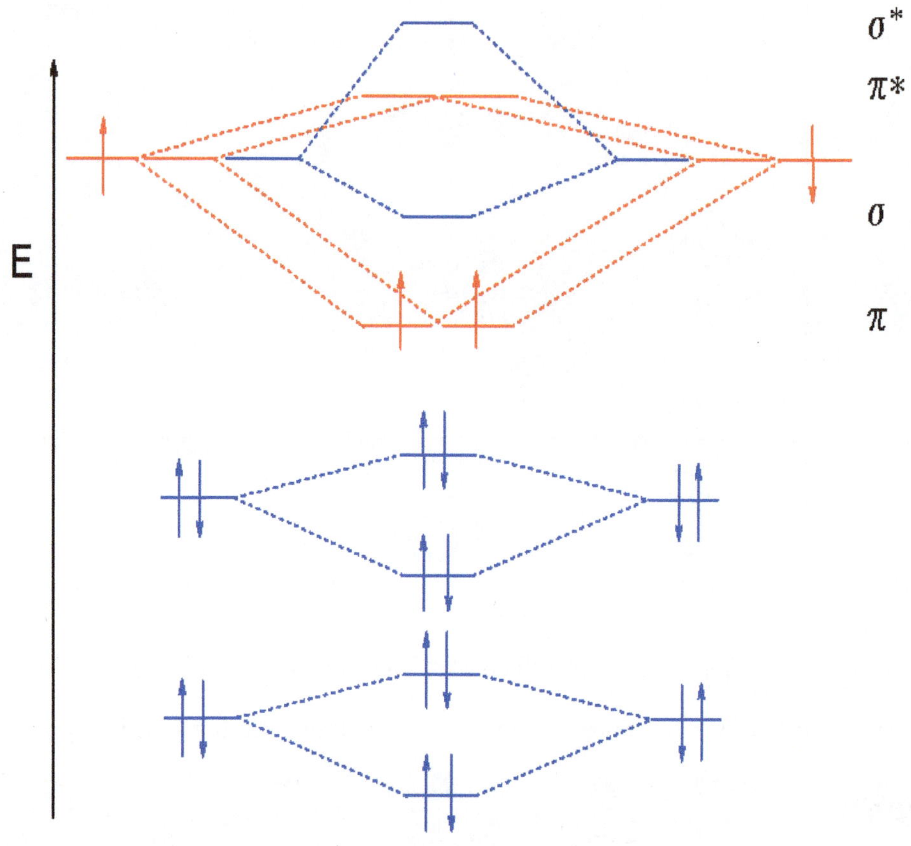

MO diagram of dioxygen triplet ground state

The bond order decreases and the bond length increases in the O_2^+ (112.2 pm), O_2 (121 pm), O_2^- (128 pm) and O_2^{2-} (149 pm).

Difluorine and Dineon

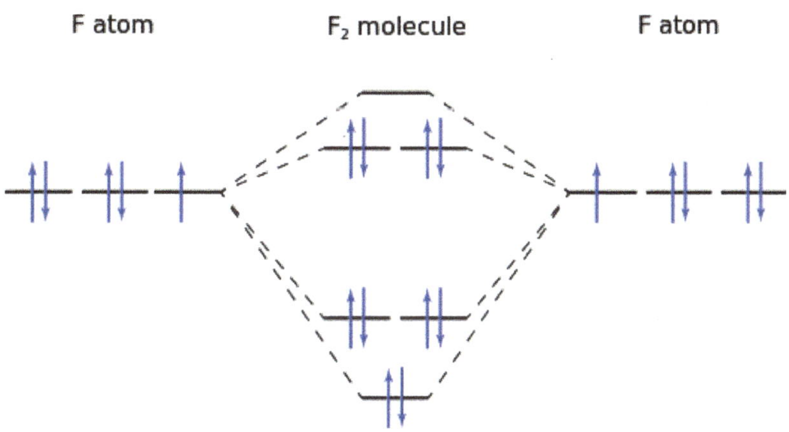

MO diagram of difluorine.

In difluorine two additional electrons occupy the 2pπ* with a bond order of 1. In dineon Ne_2 (as with dihelium) the number of bonding electrons equals the number of antibonding electrons and this compound does not exist.

Dimolybdenum and Ditungsten

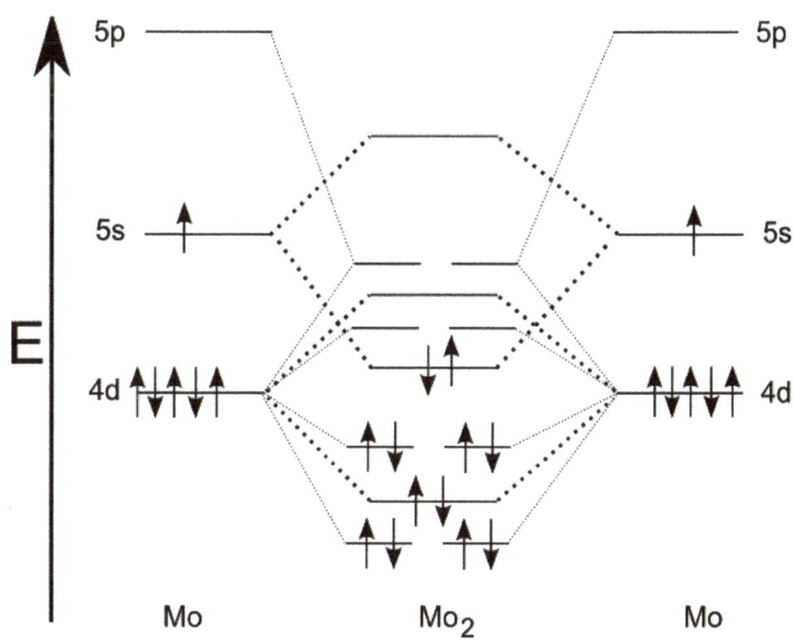

MO diagram of Dimolybdenum.

Dimolybdenum (Mo_2) is notable for having a sextuple bond. This involves two sigma bonds ($4d_z{}^2$ and 5s), two pi bonds (using $4d_{xz}$ and $4d_{yz}$), and two delta bonds ($4d_{x^2-y^2}$ and $4d_{xy}$). Ditungsten (W_2) has a similar structure.

MO energies overview

Table 1 gives an overview of MO energies for first row diatomic molecules calculated by the Hartree-Fock-Roothaan method, together with atomic orbital energies.

Table 1. Calculated MO energies for diatomic molecules in Hartrees

	H_2	Li_2	B_2	C_2	N_2	O_2	F_2
$1\sigma_g$	-0.5969	-2.4523	-7.7040	- 11.3598	- 15.6820	- 20.7296	-26.4289
$1\sigma_u$		-2.4520	-7.7032	-11.3575	-15.6783	-20.7286	-26.4286
$2\sigma_g$		-0.1816	-0.7057	-1.0613	-1.4736	-1.6488	-1.7620
$2\sigma_u$			-0.3637	-0.5172	-0.7780	-1.0987	-1.4997
$3\sigma_g$					-0.6350	-0.7358	-0.7504
$1\pi_u$			-0.3594	-0.4579	-0.6154	-0.7052	-0.8097
$1\pi_g$						-0.5319	-0.6682
1s (AO)	-0.5	-2.4778	-7.6953	-11.3255	-15.6289	-20.6686	-26.3829
2s (AO)		-0.1963	-0.4947	-0.7056	-0.9452	-1.2443	-1.5726
2p (AO)			-0.3099	-0.4333	-0.5677	-0.6319	-0.7300

Heteronuclear Diatomics

In heteronuclear diatomic molecules, mixing of atomic orbitals only occurs when the electronegativity values are similar. In carbon monoxide (CO, isoelectronic with dinitrogen) the oxygen 2s orbital is much lower in energy than the carbon 2s orbital and therefore the degree of mixing is low. The electron configuration $1\sigma^2 1\sigma^{*2} 2\sigma^2 2\sigma^{*2} 1\pi^4 3\sigma^2$ is identical to that of nitrogen. The g and u subscripts no longer apply because the molecule lacks a center of symmetry.

In hydrogen fluoride (HF), the hydrogen 1s orbital can mix with fluorine $2p_z$ orbital to form a sigma bond because experimentally the energy of 1s of hydrogen is comparable with 2p of fluorine. The HF electron configuration $1\sigma^2 2\sigma^2 3\sigma^2 1\pi^4$ reflects that the other electrons remain in three lone pairs and that the bond order is 1.

Triatomic Molecules

Carbon Dioxide

Carbon dioxide, CO_2, is a linear molecule with a total of sixteen bonding electrons in its valence shell. Carbon is the central atom of the molecule and a principal axis, the z-axis, is visualized as a single axis that goes through the center of carbon and the two oxygens atoms. For convention, blue atomic orbital lobes are positive phases, red atomic orbitals are negative phases, with respect to the wave function from the solution of the Schrödinger equation. In carbon dioxide the carbon 2s (−19.4 eV), carbon 2p (−10.7 eV), and oxygen 2p (−15.9 eV)) energies associated with the atomic orbitals are in proximity whereas the oxygen 2s energy (−32.4 eV) is different.

Carbon and each oxygen atom will have a 2s atomic orbital and a 2p atomic orbital, where the p orbital is divided into p_x, p_y, and p_z. With these derived atomic orbitals, symmetry labels are deduced with respect to rotation about the principal axis which generates a phase change, pi bond (π) or generates no phase change, known as a sigma bond (σ). Symmetry labels are further defined by whether the atomic orbital maintains its original character after an inversion about its center atom; if the atomic orbital does retain its original character it is defined gerade,g, or if the atomic orbital does not maintain its original character, ungerade, u. The final symmetry-labeled atomic orbital is now known as an irreducible representation.

Carbon dioxide's molecular orbitals are made by the linear combination of atomic orbitals of the same irreducible representation that are also similar in atomic orbital energy. Significant atomic orbital overlap explains why sp bonding may occur. Strong mixing of the oxygen 2s atomic orbital is not to be expected and are non-bonding degenerate molecular orbitals. The combination of similar atomic orbital/wave functions and the combinations of atomic orbital/wave function inverses create particular energies associated with the nonbonding (no change), bonding (lower than either parent orbital energy) and antibonding (higher energy than either parent atomic orbital energy) molecular orbitals.

- MO model carbon dioxide

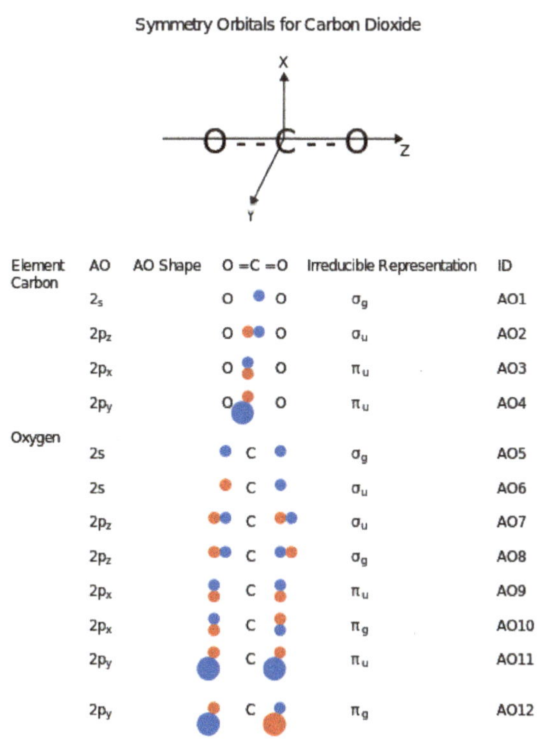

Atomic orbitals of carbon dioxide.

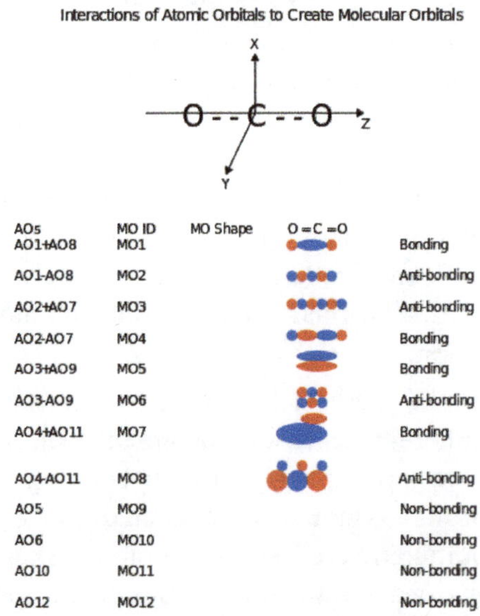

Molecular orbitals of carbon dioxide.

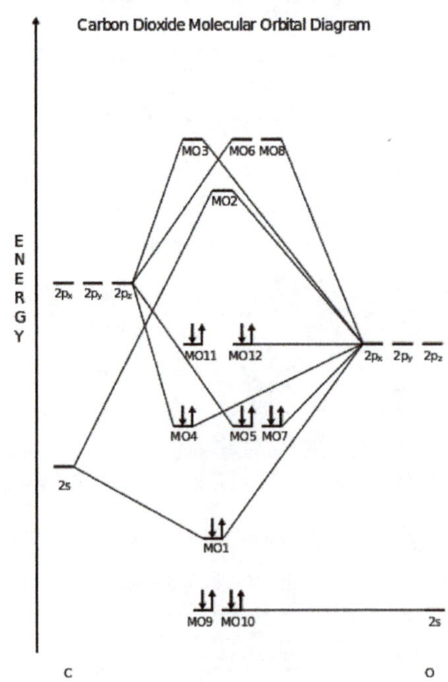

MO Diagram of carbon dioxide.

Water

Water (H_2O) is a bent molecule (105°) with C_{2v} molecular symmetry. The oxygen atomic orbitals are labeled according to their symmetry as a_1 for the 2s orbital and b_1 ($2p_x$), b_2 ($2p_y$) and a_1 ($2p_z$) for the three 2p orbitals. The two hydrogen 1s orbitals are premixed to form a_1 (σ) and b_2 (σ*) MO.

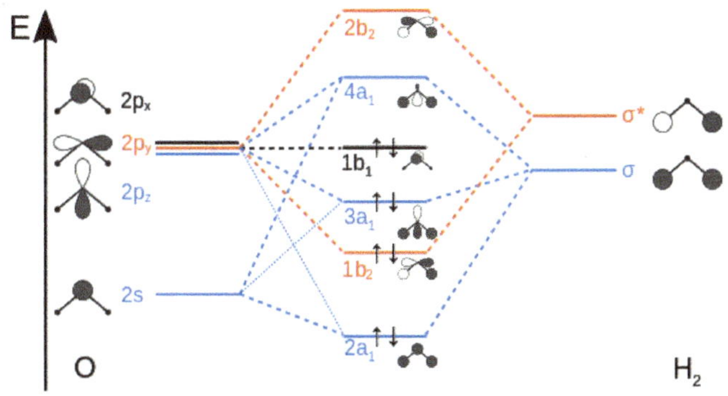

Molecular orbital diagram of water.

C_{2v}	E	C_2	$\sigma_v(xz)$	$\sigma_v'(yz)$		
A_1	1	1	1	1	z	x^2, y^2, z^2
A_2	1	1	−1	−1	R_z	xy
B_1	1	−1	1	−1	x, R_y	xz
B_2	1	−1	−1	1	y, R_x	yz

Mixing takes place between same-symmetry orbitals of comparable energy resulting a new set of MO's for water:

- $2a_1$ MO from mixing of the oxygen 2s AO and the hydrogen σ MO. Small oxygen $2p_z$ AO admixture strengthens bonding and lowers the orbital energy.
- $1b_2$ MO from mixing of the oxygen $2p_y$ AO and the hydrogen σ* MO.
- $3a_1$ MO from mixing of the oxygen $2p_z$ AO and the hydrogen σ MO. Small oxygen 2s AO admixture weakens bonding and raises the orbital energy.
- $1b_1$ nonbonding MO from the oxygen $2p_x$ AO (the p-orbital perpendicular to the molecular plane).

In agreement with this description the photoelectron spectrum for water shows a sharp peak for the nonbonding $1b_1$ MO (12.6 eV) and three broad peaks for the $3a_1$ MO (14.7 eV), $1b_2$ MO (18.5 eV) and the $2a_1$ MO (32.2 eV). The $1b_1$ MO is a lone pair, while the $3a_1$, $1b_2$ and $2a_1$ MO's can be localized to give two O–H bonds and an in-plane lone pair. This MO treatment of water does not have two equivalent rabbit ear lone pairs.

Hydrogen sulfide (H_2S) too has a C_{2v} symmetry with 8 valence electrons but the bending angle is only 92°. As reflected in its photoelectron spectrum as compared to water the $5a_1$ MO (corresponding to the $3a_1$ MO in water) is stabilised (improved overlap) and the $2b_2$ MO (corresponding to the $1b_2$ MO in water) is destabilized (poorer overlap).

Bonding in Molecular Orbitals

Orbital Degeneracy

Molecular orbitals are said to be degenerate if they have the same energy. For example, in the homonuclear diatomic molecules of the first ten elements, the molecular orbitals derived from the p_x and the p_y atomic orbitals result in two degenerate bonding orbitals (of low energy) and two degenerate antibonding orbitals (of high energy).

Ionic bonds

When the energy difference between the atomic orbitals of two atoms is quite large, one atom's orbitals contribute almost entirely to the bonding orbitals, and the others atom's orbitals contribute almost entirely to the antibonding orbitals. Thus, the situation is effectively that some electrons have been transferred from one atom to the other. This is called an (mostly) ionic bond.

Bond Order

The bond order, or number of bonds, of a molecule can be determined by combining the number of electrons in bonding and antibonding molecular orbitals. A pair of electrons in a bonding orbital creates a bond, whereas a pair of electrons in an antibonding orbital negates a bond. For example, N_2, with eight electrons in bonding orbitals and two electrons in antibonding orbitals, has a bond order of three, which constitutes a triple bond.

Bond strength is proportional to bond order—a greater amount of bonding produces a more stable bond—and bond length is inversely proportional to it—a stronger bond is shorter.

There are rare exceptions to the requirement of molecule having a positive bond order. Although Be_2 has a bond order of 0 according to MO analysis, there is experimental evidence of a highly unstable Be_2 molecule having a bond length of 245 pm and bond energy of 10 kJ/mol.

HOMO and LUMO

The highest occupied molecular orbital and lowest unoccupied molecular orbital are often referred to as the HOMO and LUMO, respectively. The difference of the energies of the HOMO and LUMO, termed the band gap, can sometimes serve as a measure of the excitability of the molecule: The smaller the energy the more easily it will be excited.

MOLECULAR ORBITAL EXAMPLES

Homonuclear Diatomics

Homonuclear diatomic MOs contain equal contributions from each atomic orbital in the basis set. This is shown in the homonuclear diatomic MO diagrams for H_2, He_2, and Li_2, all of which containing symmetric orbitals.

H_2

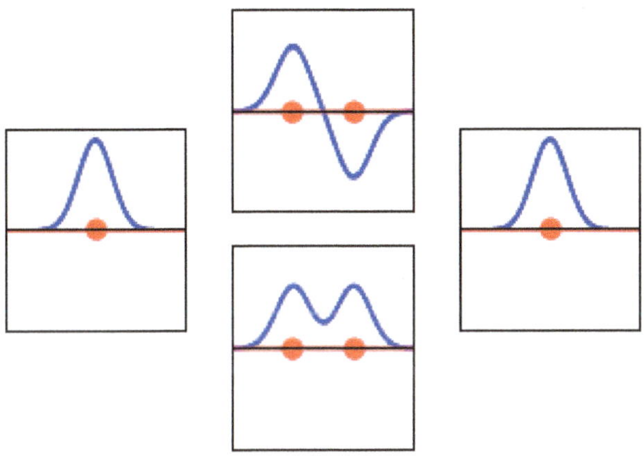

Electron wavefunctions for the 1s orbital of a lone hydrogen atom (left and right) and the corresponding bonding (bottom) and antibonding (top) molecular orbitals of the H_2 molecule. The real part of the wavefunction is the blue curve, and the imaginary part is the red curve. The red dots mark the locations of the nuclei. The electron wavefunction oscillates according to the Schrödinger wave equation, and orbitals are its standing waves. The standing wave frequency is proportional to the orbital›s kinetic energy. (This plot is a one-dimensional slice through the three-dimensional system.)

As a simple MO example consider the hydrogen molecule, H_2 (see molecular orbital diagram), with the two atoms labelled H› and H». The lowest-energy atomic orbitals, 1s› and 1s», do not transform according to the symmetries of the molecule. However, the following symmetry adapted atomic orbitals do:

1s’ – 1s”	Antisymmetric combination: negated by reflection, unchanged by other operations
1s’ + 1s”	Symmetric combination: unchanged by all symmetry operations

The symmetric combination (called a bonding orbital) is lower in energy than the basis orbitals, and the antisymmetric combination (called an antibonding orbital) is higher. Because the H_2molecule has two electrons, they can both go in the bonding orbital, making the system lower in energy (and, hence, more stable) than two free hydrogen atoms. This is called a covalent bond. The bond order is equal to the number of bonding electrons minus the number of antibonding electrons, divided by 2. In this example, there are 2 electrons in the bonding orbital and none in the antibonding orbital; the bond order is 1, and there is a single bond between the two hydrogen atoms.

He_2

On the other hand, consider the hypothetical molecule of He_2 with the atoms labeled He' and He". As with H_2, the lowest-energy atomic orbitals are the 1s' and 1s", and do not transform according to the symmetries of the molecule, while the symmetry adapted atomic orbitals do. The symmetric combination—the bonding orbital—is lower in energy than the basis orbitals, and the antisymmetric combination—the antibonding orbital—is higher. Unlike H_2, with two valence electrons, He_2 has four in its neutral ground state. Two electrons fill the lower-energy bonding orbital, σ_g(1s), while the remaining two fill the higher-energy antibonding orbital, σ_u*(1s). Thus, the resulting electron density around the molecule does not support the formation of a bond between the two atoms; without a stable bond holding the atoms together, molecule would not be expected to exist. Another way of looking at it is that there are two bonding electrons and two antibonding electrons; therefore, the bond order is 0 and no bond exists (the molecule has one bound state supported by the Van der Waals potential).

Li_2

Dilithium Li_2 is formed from the overlap of the 1s and 2s atomic orbitals (the basis set) of two Li atoms. Each Li atom contributes three electrons for bonding interactions, and the six electrons fill the three MOs of lowest energy, σ_g(1s), σ_u*(1s), and σ_g(2s). Using the equation for bond order, it is found that dilithium has a bond order of one, a single bond.

Noble Gases

Considering a hypothetical molecule of He_2, since the basis set of atomic orbitals is the same as in the case of H_2, we find that both the bonding and antibonding orbitals are filled, so there is no energy advantage to the pair. HeH would have a slight energy advantage, but not as much as H_2+ 2 He, so the molecule is very unstable and exists only briefly before decomposing into hydrogen and helium. In general, we find that atoms such as He that have full energy shells rarely bond with other atoms. Except for short-lived Van der Waals complexes, there are very few noble gas compounds known.

Heteronuclear Diatomics

While MOs for homonuclear diatomic molecules contain equal contributions from each interacting atomic orbital, MOs for heteronuclear diatomics contain different atomic orbital contributions. Orbital interactions to produce bonding or antibonding orbitals in heteronuclear diatomics occur if there is sufficient overlap between atomic orbitals as determined by their symmetries and similarity in orbital energies.

HF

In hydrogen fluoride HF overlap between the H 1s and F 2s orbitals is allowed by symmetry but the difference in energy between the two atomic orbitals prevents them from interacting to create a molecular orbital. Overlap between the H 1s and F $2p_z$ orbitals is also symmetry allowed, and these two atomic orbitals have a small energy separation. Thus, they interact, leading to creation of σ and σ* MOs and a molecule with a bond order of 1. Since HF is a non-centrosymmetric molecule, the symmetry labels g and u do not apply to its molecular orbitals.

Quantitative Approach

To obtain quantitative values for the molecular energy levels, one needs to have molecular orbitals that are such that the configuration interaction (CI) expansion converges fast towards the full CI limit. The most common method to obtain such functions is the Hartree–Fock method, which expresses the molecular orbitals as eigenfunctions of the Fock operator. One usually solves this problem by expanding the molecular orbitals as linear combinations of Gaussian functions centered on the atomic nuclei (see linear combination of atomic orbitals and basis set (chemistry)). The equation for the coefficients of these linear combinations is a generalized eigenvalue equation known as the Roothaan equations, which are in fact a particular representation of the Hartree–Fock equation. There are a number of programs in which quantum chemical calculations of MOs can be performed, including Spartan and HyperChem.

Simple accounts often suggest that experimental molecular orbital energies can be obtained by the methods of ultra-violet photoelectron spectroscopy for valence orbitals and X-ray photoelectron spectroscopy for core orbitals. This, however, is incorrect as these experiments measure the ionization energy, the difference in energy between the molecule and one of the ions resulting from the removal of one electron. Ionization energies are linked approximately to orbital energies by Koopmans› theorem. While the agreement between these two values can be close for some molecules, it can be very poor in other cases.

REFERENCES

1. Mulliken, Robert S. (July 1932). «Electronic Structures of Polyatomic Molecules and Valence. II. General Considerations». Physical Review. **41** (1): 49–71. Bibcode:1932PhRv...41...49M. doi:10.1103/PhysRev.41.49.
2. F. Hund, "Zur Deutung einiger Erscheinungen in den Molekelspektren» [On the interpretation of some phenomena in molecular spectra] Zeitschrift für Physik, vol. 36, pages 657-674 (1926).
3. F. Hund, "Zur Deutung der Molekelspektren», Zeitschrift für Physik, Part I, vol. 40, pages 742-764 (1927); Part II, vol. 42, pages 93–120 (1927); Part III, vol. 43, pages 805-826 (1927); Part IV, vol. 51, pages 759-795 (1928); Part V, vol. 63, pages 719-751 (1930).

4. R. S. Mulliken, «Electronic states. IV. Hund›s theory; second positive nitrogen and Swan bands; alternate intensities», Physical Review, vol. 29, pages 637–649 (1927).
5. R. S. Mulliken, «The assignment of quantum numbers for electrons in molecules», Physical Review, vol. 32, pages 186–222 (1928).
6. Friedrich Hund and Chemistry, Werner Kutzelnigg, on the occasion of Hund›s 100th birthday, Angewandte Chemie International Edition, 35, 573–586, (1996)
7. Robert S. Mulliken›s Nobel Lecture, Science, 157, no. 3785, 13-24. Available on-line at: Nobelprize.org
8. Sir John Lennard-Jones, "The electronic structure of some diatomic molecules", Transactions of the Faraday Society, vol. 25, pages 668-686 (1929).
9. Gary L. Miessler; Donald A. Tarr. Inorganic Chemistry. Pearson Prentice Hall, 3rd ed., 2004.
10. Catherine E. Housecroft, Alan G. Sharpe, Inorganic Chemistry, Pearson Prentice Hall; 2nd Edition, 2005, p. 29-33.
11. Peter Atkins; Julio De Paula. Atkins' Physical Chemistry. Oxford University Press, 8th ed., 2006.
12. Yves Jean; François Volatron. An Introduction to Molecular Orbitals. Oxford University Press, 1993.
13. Michael Munowitz, Principles of Chemistry, Norton & Company, 2000, p. 229-233.
14. Gagliardi, Laura; Roos, Björn O. (2005). «Quantum chemical calculations show that the uranium molecule U2 has a quintuple bond». Nature. **433**: 848–851. Bibcode:2005Natur.433..848G. doi:10.1038/nature03249.
15. Atkins., Peter ... [et al]. (2006). Inorganic chemistry (4. ed.). New York: W.H. Freeman. p. 208. ISBN 978-0-7167-4878-6.
16. Bondybey, V.E. (1984). «Electronic structure and bonding of Be2». Chemical Physics Letters. **109** (5): 436–441. doi:10.1016/0009-2614(84)80339-5.
17. Catherine E. Housecroft, Alan G, Sharpe, Inorganic Chemistry, Pearson Prentice Hall; 2nd Edition, 2005, ISBN 0130-39913-2, p. 41-43.

CHAPTER 6

Relativistic Quantum Chemistry

Chapter Contents

- **Introduction**
- **History**
- **Qualitative Treatment**
- **Periodic Table Deviations**
- **Relativistic Mechanics**

Relativistic effects are known to be very important in the physics and chemistry of the heavy elements. In fact, they are already important in some properties of light elements, but when it comes to molecules and solids containing elements of high Z, the relativistic effects become a must for essentially any property.

The basic equation in this framework is the Dirac equation, which substitutes for the Schrödinger equation. In it, the one-electron wavefunctions, called spinors, are vectors made of four components, each being a scalar function. Many computational complications appear as a consequence of this. Today, several codes exist which allow to perform Dirac-Hartre-Fock, Dirac-DFT, Dirac-HF-CI calculations in atoms and molecules of moderate size. Fortunately, approximations leading to simpler 2-component equations including spin-orbit and other spin interactions, and even to 1-component spin-free Hamiltonians, are nowadays very succesfully.

INTRODUCTION

Relativistic quantum chemistry invokes quantum chemical and relativistic mechanical arguments to explain elemental properties and structure, especially for the heavier elements of the periodic table. A prominent example of such an explanation is the color of gold; due to relativistic effects, it is not silvery like most other metals.

The term “relativistic effects” was developed in light of the history of quantum mechanics. Initially quantum mechanics was developed without considering the theory of relativity. By convention, “relativistic effects” are those discrepancies between values calculated by models considering and not considering relativity. Relativistic effects are important for the heavier elements with high atomic numbers. In the most common layout of the periodic table, these elements are shown in the lower area. Examples are the lanthanides and actinides.

Relativistic effects in chemistry can be considered to be perturbations, or small corrections, to the non-relativistic theory of chemistry, which is developed from the solutions of the Schrödinger equation. These corrections affect the electrons differently depending on the electron speed relative to the speed of light. Relativistic effects are more prominent in heavy elements because only in these elements do electrons attain more pronounced relativistic speeds.

HISTORY

Beginning in 1935, Bertha Swirles described a relativistic treatment of a many-electron system, in spite of Paul Dirac’s 1929 assertion that the only imperfections remaining in quantum mechanics “...give rise to difficulties only when high-speed particles are involved, and are therefore of no importance in the consideration of atomic and molecular structure and ordinary chemical reactions in which it is, indeed, usually sufficiently accurate if one neglects relativity variation of mass and velocity and assumes only Coulomb forces between the various electrons and atomic nuclei.”

Theoretical chemists by and large agreed with Dirac’s sentiment until the 1970s, when relativistic effects began to become realized in heavy elements. The Schrödinger equation had been developed without considering relativity in Schrödinger’s 1926 paper. Relativistic corrections were made to the Schrödinger equation (see Klein–Gordon equation) in order to explain the fine structure of atomic spectra, but this development and others did not immediately trickle into the chemical community. Since atomic spectral lines were largely in the realm of physics and not in that of chemistry, most chemists were unfamiliar with relativistic quantum mechanics, and their attention was on lighter elements typical for the organic chemistry focus of the time.

Dirac’s opinion on the role relativistic quantum mechanics would play for chemical systems is wrong for two reasons: the first being that electrons in s and p atomic orbitals travel at a significant fraction of the speed of light and the second being that there are indirect consequences of relativistic effects which are especially evident for d and f atomic orbitals.

QUALITATIVE TREATMENT

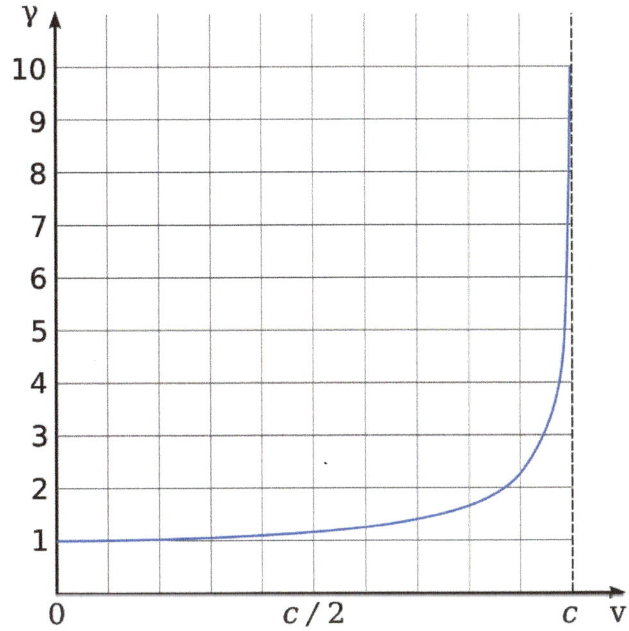

Relativistic γ as a function of velocity. For a small velocity, the E_{rel} (ordinate) is equal to $E_0 = mc^2$ but as $v_e \to c$ the E_{rel} goes to infinity.

One of the most important and familiar results of relativity is that the relativistic mass of the electron increases by

$$m_{rel} = \frac{m_e}{\sqrt{1-(v_e/c)^2}}$$

where m_e, v_e, c are the electron rest mass, velocity of the electron, and speed of light respectively. The figure at the right illustrates the relativistic effects on the mass of an electron as a function of its velocity.

This has an immediate implication on the Bohr radius (a_0) which is given by

$$a_0 = \frac{\hbar}{m_e c\alpha}$$

where $\hbar$ is the reduced Planck's constant and α is the fine-structure constant (a relativistic correction for the Bohr model).

Arnold Sommerfeld calculated that, for a 1s electron of a hydrogen atom with an orbiting radius of 0.0529 nm, $\alpha \approx 1/137$. That is to say, the fine-structure constant shows the electron traveling at nearly 1/137 the speed of light. One can extend this to a larger element by using the expression $v \approx Zc/137$ for a 1s electron where v is its radial velocity. For gold with ($Z = 79$) the 1s electron will be going ($\alpha = 0.58c$) 58% of the speed of light. Plugging this in for v/c for the relativistic mass one finds that $m_{rel} = 1.22m_e$ and in turn putting this in for the Bohr radius above one finds that the radius shrinks by 22%.

If one substitutes in the relativistic mass into the equation for the Bohr radius it can be written

$$a_{rel} = \frac{\hbar\sqrt{1-(v_e/c)^2}}{m_e c\alpha}$$

It follows that

$$\frac{a_{rel}}{a_0} = \sqrt{1-(v_e/c)^2}$$

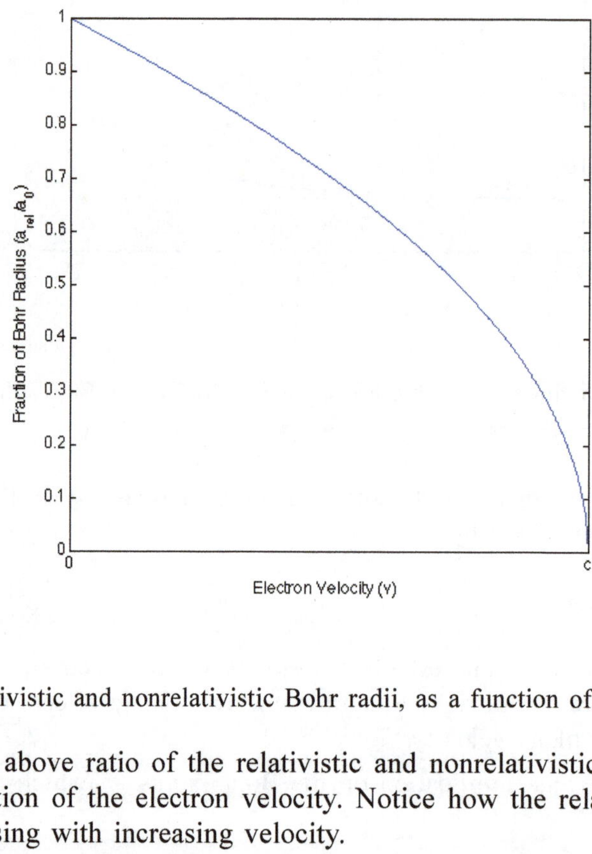

Ratio of relativistic and nonrelativistic Bohr radii, as a function of electron velocity.

At right, the above ratio of the relativistic and nonrelativistic Bohr radii has been plotted as a function of the electron velocity. Notice how the relativistic model shows the radius decreasing with increasing velocity.

When the Bohr treatment is extended to hydrogenic-like atoms using the Quantum Rule, the Bohr radius becomes

$$r = \frac{n^2\hbar^2 4\pi\epsilon_0}{m_e Ze^2}$$

where n is the principal quantum number and Z is an integer for the atomic number. From quantum mechanics the angular momentum is given as $mv_e r = n\hbar$. Substituting into the equation above and solving for v gives

$$r = \frac{mv_e rn\hbar 4\pi\epsilon_0}{mZe^2}$$

$$1 = \frac{v_e n\hbar 4\pi\epsilon_0}{Ze^2}$$

$$v_e = \frac{Ze^2}{n\hbar 4\pi\epsilon_0}$$

From this point atomic units can be used to simplify the expression into

$$v_e = \frac{Z}{n}$$

Substituting this into the expression for the Bohr ratio mentioned above gives

$$\frac{a_{rel}}{a_0} = \sqrt{1 - \left(\frac{Z}{nc}\right)^2}$$

At this point onc can scc that for a low value of n and a high value of Z that $\frac{a_{rel}}{a_0} < 1$. This fits with intuition: electrons with lower principal quantum numbers will have a higher probability density of being nearer to the nucleus. A nucleus with a large charge will cause an electron to have a high velocity. A higher electron velocity means an increased electron relativistic mass, as a result the electrons will be near the nucleus more of the time and thereby contract the radius for small principal quantum numbers.

PERIODIC TABLE DEVIATIONS

The periodic table was constructed by scientists who noticed periodic trends in known elements of the time. Indeed, the patterns found in it is what gives the periodic table its power. Many of the chemical and physical differences between the 6th period (Cs–Rn) and the 5th period (Rb–Xe) arise from the larger relativistic effects for the former. These relativistic effects are particularly large for gold and its neighbors, platinum and mercury.

Mercury

Mercury (Hg) is a liquid down to −39°C (see Melting Point (m.p.)). Bonding forces are weaker for Hg–Hg bonds than for its immediate neighbors such as cadmium (m.p. 321°C) and gold (m.p. 1064°C). The lanthanide contraction is a partial explanation; however, it does not entirely account for this anomaly. In the gas phase mercury is alone in metals in that it is quite typically found in a monomeric form as Hg(g). Hg_2^{2+}(g) also forms and it is a stable species due to the relativistic shortening of the bond.

$Hg_2(g)$ does not form because the $6s^2$ orbital is contracted by relativistic effects and may therefore only weakly contribute to any bonding; in fact Hg–Hg bonding must be mostly the result of van der Waals forces, which explains why the bonding for Hg–Hg is weak enough to allow for Hg to be a liquid at room temperature.

$Au_2(g)$ and Hg(g) are analogous, at the least in having the same nature of difference, to $H_2(g)$ and He(g). It is for the relativistic contraction of the $6s^2$ orbital that gaseous mercury can be called a pseudo noble gas.

Color of Gold and Caesium

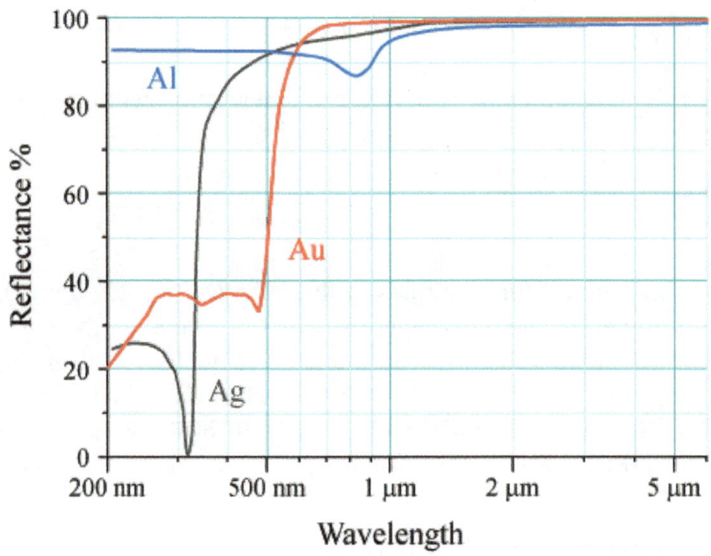

Spectral reflectance curves for aluminum (Al), silver (Ag), and gold (Au) metal mirrors.

The reflectivity of Au, Ag, Al is shown on the figure to the right. The human eye sees electromagnetic radiation with a wavelength near 600 nm as yellow. As is clear from its reflectance spectrum, gold appears yellow because it absorbs blue light more than it absorbs other visible wavelengths of light; the reflected light (which is what we see) is therefore lacking in blue compared to the incident light. Since yellow is complementary to blue, this makes a piece of gold appear yellow (under white light) to human eyes.

The electronic transition responsible for this absorption is a transition from the 5d to the 6s level. An analogous transition occurs in Ag but the relativistic effects are lower in Ag so while the 4d experiences some expansion and the 5s some contraction, the 4d-5s distance in Ag is still much greater than the 5d-6s distance in Au because the relativistic effects in Ag are smaller than those in Au. Thus, non-relativistic gold would be white. The relativistic effects are raising the 5d orbital and lowering the 6s orbital.

A similar effect occurs in caesium metal, the heaviest of the alkali metals which can be collected in quantities sufficient to allow viewing. Whereas the other alkali metals are silver-white, caesium metal has a distinctly golden hue.

Inert Pair Effect

In Tl(I) (thallium), Pb(II) (lead), and Bi(III) (bismuth) complexes there is a $6s^2$ electron pair. The 'inert pair effect' refers to the tendency for this pair of electrons to resist oxidation due to a relativistic contraction of the 6s orbital.

Others

Some of the phenomena commonly attributed to relativistic effects are:

- The existence of mercury(IV) fluoride
- Aurophilicity
- The stability of the gold anion, Au^-, in compounds such as CsAu
- The crystal structure of lead, which is face-centered cubic instead of diamond-like
- The striking similarity between zirconium and hafnium
- The stability of the uranyl cation, as well as other high oxidation states in the early actinides (Pa-Am)
- The small atomic radii of francium and radium
- About 10% of the lanthanide contraction is attributed to the relativistic mass of high velocity electrons and the smaller Bohr radius that results.
- In the case of gold, a lot more than 10% of its contraction is due to relativistically heavy electrons, and gold #79 is almost twice as dense as lead #82.

RELATIVISTIC MECHANICS

In physics, **relativistic mechanics** refers to mechanics compatible with special relativity (SR) and general relativity (GR). It provides a non-quantum mechanical description of a system of particles, or of a fluid, in cases where the velocities of moving objects are comparable to the speed of light *c*. As a result, classical mechanics is extended correctly to particles traveling at high velocities and energies, and provides a consistent inclusion of electromagnetism with the mechanics of particles. This was not possible in Galilean relativity, where it would be permitted for particles and light to travel at *any* speed, including faster than light. The foundations of relativistic mechanics are the postulates of special relativity and general relativity. The unification of SR with quantum mechanics is relativistic quantum mechanics, while attempts for that of GR is quantum gravity, an unsolved problem in physics.

As with classical mechanics, the subject can be divided into "kinematics"; the description of motion by specifying positions, velocities and accelerations, and "dynamics"; a full description by considering energies, momenta, and angular momenta and their conservation laws, and forces acting on particles or exerted by particles. There is however a subtlety; what appears to be "moving" and what is "at rest"—which is termed by "statics" in classical mechanics—depends on the relative motion of observers who measure in frames of reference.

Although some definitions and concepts from classical mechanics do carry over to SR, such as force as the time derivative of momentum (Newton's second law), the work done by a particle as the line integral of force exerted on the particle along a path, and power as the time derivative of work done, there are a number of significant modifications to the remaining definitions and formulae. SR states that motion is relative and the laws of physics are the same for all experimenters irrespective of their inertial reference frames. In addition to modifying notions of space and time, SR forces one to reconsider the concepts of mass, momentum, and energy all of which are important constructs in Newtonian mechanics. SR shows that these concepts are all different aspects of the same physical quantity in much the same way that it shows space and time to be interrelated. Consequently, another modification is the concept of the center of mass of a system, which is straightforward to define in classical mechanics but much less obvious in relativity - see relativistic center of mass for details.

The equations become more complicated in the more familiar three-dimensional vector calculus formalism, due to the nonlinearity in the Lorentz factor, which accurately accounts for relativistic velocity dependence and the speed limit of all particles and fields. However, they have a simpler and elegant form in *four*-dimensional spacetime, which includes flat Minkowski space (SR) and curved spacetime (GR), because three-dimensional vectors derived from space and scalars derived from time can be collected into four vectors, or four-dimensional tensors. However, the six component angular momentum tensor is sometimes called a bivector because in the 3D viewpoint it is two vectors (one of these, the conventional angular momentum, being an axial vector).

Relativistic Kinematics

The relativistic four-velocity, that is the four-vector representing velocity in relativity, is defined as follows:

$$\mathbf{U} = \frac{d\mathbf{X}}{d\tau} = \left(\frac{cdt}{d\tau}, \frac{d\mathbf{x}}{d\tau}\right)$$

In the above, τ is the proper time of the path through spacetime, called the world-line, followed by the object velocity the above represents, and

$$\mathbf{X} = (ct, \mathbf{x})$$

is the four-position; the coordinates of an event. Due to time dilation, the proper time is the time between two events in a frame of reference where they take place at the same location. The proper time is related to coordinate time t by:

$$\frac{d\tau}{dt} = \frac{1}{\gamma(\mathbf{v})}$$

where $\gamma(\mathbf{v})$ is the Lorentz factor:

$$\gamma(\mathbf{v}) = \frac{1}{\sqrt{1 - \mathbf{v} \cdot \mathbf{v}/c^2}} \rightleftharpoons \gamma(v) = \frac{1}{\sqrt{1 - (v/c)^2}}.$$

(either version may be quoted) so it follows:

$$\mathbf{U} = \gamma(\mathbf{v})(c, \mathbf{v})$$

The first three terms, excepting the factor of $\gamma(\mathbf{v})$, is the velocity as seen by the observer in their own reference frame. The $\gamma(\mathbf{v})$ is determined by the velocity $\mathbf{v}$ between the observer's reference frame and the object's frame, which is the frame in which its proper time is measured. This quantity is invariant under Lorentz transformation, so to check to see what an observer in a different reference frame sees, one simply multiplies the velocity four-vector by the Lorentz transformation matrix between the two reference frames.

Relativistic Dynamics

Relativistic Energy and Momentum

There are a couple of (equivalent) ways to define momentum and energy in SR. One method uses conservation laws. If these laws are to remain valid in SR they must be true in every possible reference frame. However, if one does some simple thought experiments using the Newtonian definitions of momentum and energy, one sees that these quantities are not conserved in SR. One can rescue the idea of conservation by making some small modifications to the definitions to account for relativistic velocities. It is these new definitions which are taken as the correct ones for momentum and energy in SR.

The four-momentum of an object is straightforward, identical in form to the classical momentum, but replacing 3-vectors with 4-vectors:

$$\mathbf{P} = m_0\mathbf{U} = (E/c, \mathbf{p})$$

The energy and momentum of an object with invariant mass m_0 (also called *rest mass*), moving with velocity $\mathbf{v}$ with respect to a given frame of reference, are respectively given by

$$E = \gamma(\mathbf{v})m_0c^2$$
$$\mathbf{p} = \gamma(\mathbf{v})m_0\mathbf{v}$$

The factor of $\gamma(\mathbf{v})$ comes from the definition of the four-velocity described above. The appearance of the γ factor has an alternative way of being stated, explained in the next section.

The kinetic energy, K, is defined as

$$K = (\gamma - 1)m_0c^2 = E - m_0c^2\,,$$

And the speed as a function of kinetic energy is given by

$$v = \frac{c\sqrt{K(K + 2m_0c^2)}}{K + m_0c^2} = \frac{c\sqrt{(E - m_0c^2)(E + m_0c^2)}}{E} = \frac{pc^2}{E}.$$

Rest Mass and Relativistic Mass

The quantity

$$m = \gamma(\mathbf{v})m_0$$

is often called the *relativistic mass* of the object in the given frame of reference.

This makes the relativistic relation between the spatial velocity and the spatial momentum look identical. However, this can be misleading, as it is *not* appropriate in special relativity in *all* circumstances. For instance, kinetic energy and force in special relativity can not be written exactly like their classical analogues by only replacing the mass with the relativistic mass. Moreover, under Lorentz transformations, this relativistic mass is not invariant, while the rest mass is. For this reason many people find it easier use the rest mass (thereby introduce γ through the 4-velocity or coordinate time), and discard the concept of relativistic mass.

Lev B. Okun suggested that "this terminology ... has no rational justification today", and should no longer be taught.

Other physicists, including Wolfgang Rindler and T. R. Sandin, have argued that relativistic mass is a useful concept and there is little reason to stop using it. See mass in special relativity for more information on this debate.

Some authors use m for relativistic mass and m_0 for rest mass, others simply use m for rest mass. This article uses the former convention for clarity.

The energy and momentum of an object with invariant mass m_0 are related by the formulas

$$E^2 - (pc)^2 = (m_0c^2)^2$$
$$\mathbf{p}c^2 = E\mathbf{v}\,.$$

The first is referred to as the *relativistic energy–momentum relation*. It can be derived by considering that $\frac{v^2}{c^2}$ can be written as $\frac{p^2}{\gamma^2 m_0^2 c^2}$ where the denominator can be written as $\frac{E^2}{c^2}$. Now, gamma can be replaced in the expression of energy. While the energy E and the momentum $\mathbf{p}$ depend on the frame of reference in which they are measured, the quantity $E^2 - (pc)^2$ is invariant, and arises as $-c^2$ times the squared magnitude of the 4-momentum vector which is $-(m_0c)^2$.

It should be noted that the invariant mass of a system

$$m_{0\,\mathrm{tot}} = \frac{\sqrt{E_{\mathrm{tot}}^2 - (p_{\mathrm{tot}}c)^2}}{c^2}$$

is different from the sum of the rest masses of the particles of which it is composed due to kinetic energy and binding energy. Rest mass is not a conserved quantity in special relativity unlike the situation in Newtonian physics. However, if an object is not changing internally, then its rest mass will not change and can be calculated with the same result in any frame of reference.

A particle whose rest mass is zero is called *massless*. Photons and gravitons are thought to be massless; and neutrinos are nearly so.

Mass–energy Equivalence

The relativistic energy–momentum equation holds for all particles, even for massless particles for which $m_0 = 0$. In this case:

$$E = pc$$

When substituted into $Ev = c^2p$, this gives $v = c$: massless particles (such as photons) always travel at the speed of light.

Notice that the rest mass of a composite system will generally be slightly different from the sum of the rest masses of its parts since, in its rest frame, their kinetic energy will increase its mass and their (negative) binding energy will decrease its mass. In particular, a hypothetical "box of light" would have rest mass even though made of particles which do not since their momenta would cancel.

Looking at the above formula for invariant mass of a system, one sees that, when a single massive object is at rest ($\mathbf{v} = \mathbf{0}$, $\mathbf{p} = \mathbf{0}$), there is a non-zero mass remaining: $m_0 = E/c^2$. The corresponding energy, which is also the total energy when a single particle is at rest, is referred to as "rest energy". In systems of particles which are seen from a moving inertial frame, total energy increases and so does momentum. However, for single particles the rest mass remains constant, and for systems of particles the invariant mass remain constant, because in both cases, the energy and momentum increases subtract from each other, and cancel. Thus, the invariant mass of systems of particles is a calculated constant for all observers, as is the rest mass of single particles.

The mass of Systems and Conservation of Invariant Mass

For systems of particles, the energy–momentum equation requires summing the momentum vectors of the particles:

$$E^2 - \mathbf{p} \cdot \mathbf{p}c^2 = m_0^2 c^4$$

The inertial frame in which the momenta of all particles sums to zero is called the center of momentum frame. In this special frame, the relativistic energy–momentum equation has $\mathbf{p} = 0$, and thus gives the invariant mass of the system as merely the total energy of all parts of the system, divided by c^2

$$m_{0,\,\text{system}} = \sum_n E_n / c^2$$

This is the invariant mass of any system which is measured in a frame where it has zero total momentum, such as a bottle of hot gas on a scale. In such a system, the mass which the scale weighs is the invariant mass, and it depends on the total energy of the system. It is thus more than the sum of the rest masses of the molecules, but also includes all the totaled energies in the system as well. Like energy and momentum, the invariant mass of isolated systems cannot be changed so long as the system remains totally closed (no mass or energy allowed in or out), because the total relativistic energy of the system remains constant so long as nothing can enter or leave it.

An increase in the energy of such a system which is caused by translating the system to an inertial frame which is not the center of momentum frame, causes an increase in energy and momentum without an increase in invariant mass. $E = m_0c^2$, however, applies only to isolated systems in their center-of-momentum frame where momentum sums to zero.

Taking this formula at face value, we see that in relativity, mass is simply energy by another name (and measured in different units). In 1927 Einstein remarked about special relativity, "Under this theory mass is not an unalterable magnitude, but a magnitude dependent on (and, indeed, identical with) the amount of energy."

Closed (isolated) Systems

In a "totally-closed" system (i.e., isolated system) the total energy, the total momentum, and hence the total invariant mass are conserved. Einstein's formula for change in mass translates to its simplest $\Delta E = \Delta mc^2$ form, however, only in non-closed systems in which energy is allowed to escape (for example, as heat and light), and thus invariant mass is reduced. Einstein's equation shows that such systems must lose mass, in accordance with the above formula, in proportion to the energy they lose to the surroundings. Conversely, if one can measure the differences in mass between a system before it undergoes a reaction which releases heat and light, and the system after the reaction when heat and light have escaped, one can estimate the amount of energy which escapes the system.

Chemical and Nuclear Reactions

In both nuclear and chemical reactions, such energy represents the difference in binding energies of electrons in atoms (for chemistry) or between nucleons in nuclei (in atomic reactions). In both cases, the mass difference between reactants and (cooled) products measures the mass of heat and light which will escape the reaction, and thus (using the equation) give the equivalent energy of heat and light which may be emitted if the reaction proceeds.

In chemistry, the mass differences associated with the emitted energy are around 10^{-9} of the molecular mass. However, in nuclear reactions the energies are so large that they are associated with mass differences, which can be estimated in advance, if the products and reactants have been weighed (atoms can be weighed indirectly by using atomic masses, which are always the same for each nuclide). Thus, Einstein's formula becomes important when one has measured the masses of different atomic nuclei. By looking at the difference in masses, one can predict which nuclei have stored energy that can be released by certain nuclear reactions, providing important information which was useful in the development of nuclear energy and, consequently, the nuclear bomb. Historically, for example, Lise Meitner was able to use the mass differences in nuclei to estimate that there was enough energy available to make nuclear fission a favorable process. The implications of this special form of Einstein's formula have thus made it one of the most famous equations in all of science.

Center of Momentum Frame

The equation $E = m_0c^2$ applies only to isolated systems in their center of momentum frame. It has been popularly misunderstood to mean that mass may be *converted* to energy, after which the *mass* disappears. However, popular explanations of the equation as applied to systems include open (non-isolated) systems for which heat and light are allowed to escape, when they otherwise would have contributed to the mass (invariant mass) of the system.

Historically, confusion about mass being "converted" to energy has been aided by confusion between mass and "matter", where matter is defined as fermion particles. In such a definition, electromagnetic radiation and kinetic energy (or heat) are not considered "matter". In some situations, matter may indeed be converted to non-matter forms of energy (see above), but in all these situations, the matter and non-matter forms of energy still retain their original mass.

For isolated systems (closed to all mass and energy exchange), mass never disappears in the center of momentum frame, because energy cannot disappear. Instead, this equation, in context, means only that when any energy is added to, or escapes from, a system in the center-of-momentum frame, the system will be measured as having gained or lost mass, in proportion to energy added or removed. Thus, in theory, if an atomic bomb were placed in a box strong enough to hold its blast, and detonated upon a scale, the mass of this closed system would not change, and the scale would not move. Only when a transparent "window" was opened in the super-strong plasma-filled box, and light and heat were allowed to escape in a beam, and the bomb components to cool, would the system lose the mass associated with the energy of the blast. In a 21 kiloton bomb, for example, about a gram of light and heat is created. If this heat and light were allowed to escape, the remains of the bomb would lose a gram of mass, as it cooled. In this thought-experiment, the light and heat carry away the gram of mass, and would therefore deposit this gram of mass in the objects that absorb them.

Angular momentum

In relativistic mechanics, the time-varying mass moment

$$\mathbf{N} = m(\mathbf{x} - t\mathbf{v})$$

and orbital 3-angular momentum

$$\mathbf{L} = \mathbf{x} \times \mathbf{p}$$

of a point-like particle are combined into a four-dimensional bivector in terms of the 4-position **X** and the 4-momentum **P** of the particle:

$$\mathbf{M} = \mathbf{X} \wedge \mathbf{P}$$

where $\wedge$ denotes the exterior product. This tensor is additive: the total angular momentum of a system is the sum of the angular momentum tensors for each constituent of the system. So, for an assembly of discrete particles one sums the angular momentum tensors over the particles, or integrates the density of angular momentum over the extent of a continuous mass distribution.

Each of the six components forms a conserved quantity when aggregated with the corresponding components for other objects and fields.

Force

In special relativity, Newton's second law does not hold in the form $\mathbf{F} = m\mathbf{a}$, but it does if it is expressed as

$$\mathbf{F} = \frac{d\mathbf{p}}{dt}$$

where $\mathbf{p} = \gamma(\mathbf{v})m_0\mathbf{v}$ is the momentum as defined above and m_0 is the invariant mass. Thus, the force is given by

$$\mathbf{F} = \gamma(\mathbf{v})^3 m_0 \, \mathbf{a}_\parallel + \gamma(\mathbf{v}) m_0 \, \mathbf{a}_\perp$$

Consequently, in some old texts, $\gamma(\mathbf{v})^3 m_0$ is referred to as the *longitudinal mass*, and $\gamma(\mathbf{v})m_0$ is referred to as the *transverse mass*, which is numerically the same as the relativistic mass. See mass in special relativity.

If one inverts this to calculate acceleration from force, one gets

$$\mathbf{a} = \frac{1}{m_0\gamma(\mathbf{v})}\left(\mathbf{F} - \frac{(\mathbf{v}\cdot\mathbf{F})\mathbf{v}}{c^2}\right)$$

The force described in this section is the classical 3-D force which is not a four-vector. This 3-D force is the appropriate concept of force since it is the force which obeys Newton's third law of motion. It should not be confused with the so-called four-force which is merely the 3-D force in the comoving frame of the object transformed as if it were a four-vector. However, the density of 3-D force (linear momentum transferred per unit four-volume) *is* a four-vector (density of weight +1) when combined with the negative of the density of power transferred.

Torque

The torque acting on a point-like particle is defined as the derivative of the angular momentum tensor given above with respect to proper time:

$$\mathbf{\Gamma} = \frac{d\mathbf{M}}{d\tau} = \mathbf{X} \wedge \mathbf{F}$$

or in tensor components:

$$\Gamma_{\alpha\beta} = X_\alpha F_\beta - X_\beta F_\alpha$$

where **F** is the 4d force acting on the particle at the event **X**. As with angular momentum, torque is additive, so for an extended object one sums or integrates over the distribution of mass.

Kinetic Energy

The *work-energy theorem* says the change in kinetic energy is equal to the work done on the body. In special relativity:

$$\Delta K = W = [\gamma_1 - \gamma_0]m_0c^2.$$

If in the initial state the body was at rest, so $v_0 = 0$ and $\gamma_0(v_0) = 1$, and in the final state it has speed $v_1 = v$, setting $\gamma_1(v_1) = \gamma(v)$, the kinetic energy is then;

$$K = [\gamma(v) - 1]m_0c^2,$$

a result that can be directly obtained by subtracting the rest energy m_0c^2 from the total relativistic energy $\gamma(v)m_0c^2$.

Classical limit

The Lorentz factor $\gamma(v)$ can be expanded into a Taylor series or binomial series for $(v/c)^2 < 1$, obtaining:

$$\gamma = \frac{1}{\sqrt{1-(v/c)^2}} = \sum_{n=0}^{\infty}\left(\frac{v}{c}\right)^{2n}\prod_{k=1}^{n}\left(\frac{2k-1}{2k}\right) = 1 + \frac{1}{2}\left(\frac{v}{c}\right)^2 + \frac{3}{8}\left(\frac{v}{c}\right)^4 + \frac{5}{16}\left(\frac{v}{c}\right)^6 + \cdots$$

and consequently

$$E - m_0c^2 = \frac{1}{2}m_0v^2 + \frac{3}{8}\frac{m_0v^4}{c^2} + \frac{5}{16}\frac{m_0v^6}{c^4} + \cdots;$$

$$\mathbf{p} = m_0\mathbf{v} + \frac{1}{2}\frac{m_0v^2\mathbf{v}}{c^2} + \frac{3}{8}\frac{m_0v^4\mathbf{v}}{c^4} + \frac{5}{16}\frac{m_0v^6\mathbf{v}}{c^6} + \cdots.$$

For velocities much smaller than that of light, one can neglect the terms with c^2 and higher in the denominator. These formulas then reduce to the standard definitions of Newtonian kinetic energy and momentum. This is as it should be, for special relativity must agree with Newtonian mechanics at low velocities.

REFERENCES

1. Mulliken, Robert S. (July 1932). "Electronic Structures of Polyatomic Molecules and Valence. II. General Considerations". Physical Review. **41** *(1): 49–71. Bibcode*:1932PhRv...41...49M. doi:10.1103/PhysRev.41.49.
2. F. Hund, "Zur Deutung einiger Erscheinungen in den Molekelspektren" [On the interpretation of some phenomena in molecular spectra] Zeitschrift für Physik, vol. 36, pages 657-674 (1926).
3. F. Hund, "Zur Deutung der Molekelspektren", Zeitschrift für Physik, Part I, vol. 40, pages 742-764 (1927); Part II, vol. 42, pages 93–120 (1927); Part III, vol. 43, pages 805-826 (1927); Part IV, vol. 51, pages 759-795 (1928); Part V, vol. 63, pages 719-751 (1930).

4. R. S. Mulliken, "Electronic states. IV. Hund's theory; second positive nitrogen and Swan bands; alternate intensities", Physical Review, vol. 29, pages 637–649 (1927).
5. R. S. Mulliken, "The assignment of quantum numbers for electrons in molecules", Physical Review, vol. 32, pages 186–222 (1928).
6. Friedrich Hund and Chemistry, Werner Kutzelnigg, on the occasion of Hund's 100th birthday, Angewandte Chemie International Edition, 35, 573–586, (1996)
7. Robert S. Mulliken's Nobel Lecture, Science, 157, no. 3785, 13-24. Available on-line at: Nobelprize.org
8. Sir John Lennard-Jones, "The electronic structure of some diatomic molecules", Transactions of the Faraday Society, vol. 25, pages 668-686 (1929).
9. Gary L. Miessler; Donald A. Tarr. Inorganic Chemistry. Pearson Prentice Hall, 3rd ed., 2004.
10. Catherine E. Housecroft, Alan G. Sharpe, Inorganic Chemistry, Pearson Prentice Hall; 2nd Edition, 2005, p. 29-33.
11. Peter Atkins; Julio De Paula. Atkins' Physical Chemistry. Oxford University Press, 8th ed., 2006.
12. Yves Jean; François Volatron. An Introduction to Molecular Orbitals. Oxford University Press, 1993.
13. Michael Munowitz, Principles of Chemistry, Norton & Company, 2000, p. 229-233.
14. Gagliardi, Laura; Roos, Björn O. (2005). "Quantum chemical calculations show that the uranium molecule U2 has a quintuple bond". Nature. **433**: 848–851. Bibcode:2005Natur.433..848G. doi:10.1038/nature03249.
15. Atkins., Peter ... [et al]. (2006). Inorganic chemistry (4. ed.). New York: W.H. Freeman. p. 208. ISBN 978-0-7167-4878-6.
16. Bondybey, V.E. (1984). "Electronic structure and bonding of Be2". Chemical Physics Letters. **109** *(5): 436–441.* doi:10.1016/0009-2614(84)80339-5.
17. Catherine E. Housecroft, Alan G, Sharpe, Inorganic Chemistry, Pearson Prentice Hall; 2nd Edition, 2005, ISBN 0130-39913-2, p. 41-43.

CHAPTER 7

Schrödinger Equation

Chapter Contents

- **Introduction**
- **Equation**
- **Implications**
- **Interpretation of the Wave Function**
- **Historical Background and Development**
- **The Wave Equation for Particles**
- **Nonrelativistic Quantum Mechanics**
- **Properties**
- **Relativistic Quantum Mechanics**
- **Quantum Field Theory**
- **First Order Form**

The Shrodinger equation is:

$$\hbar\omega = (\hbar^2 k^2 / 2m_0) + V$$

$$\Psi(x,t) = A\exp(-i(\omega t - kx))$$

$$d\Psi / dt = -i\omega\Psi \quad \text{and} \quad d^2\Psi / dx^2 = -k^2\Psi$$

$$i\hbar\frac{d\Psi}{dt} = \left(-\frac{\hbar^2}{2m_0}\frac{d^2\Psi}{dx^2}\right) + V\Psi$$

The solution to this equation is a wave that describes the quantum aspects of a system. However, physically interpreting the wave is one of the main philosophical problems of quantum mechanics.

The solution to the equation is based on the method of Eigen Values devised by Fourier. This is where any mathematical function is expressed as the sum of an infinite series of other periodic functions. The trick is to find the correct functions that have the right amplitudes so that when added together by superposition they give the desired solution.

So, the solution to Schrondinger's equation, the wave function for the system, was replaced by the wave functions of the individual series, natural harmonics of each other, an infinite series. Shrodinger has discovered that the replacement waves described the individual states of the quantum system and their amplitudes gave the relative importance of that state to the whole system.

Schrodinger's equation shows all of the wave like properties of matter and was one of greatest achievements of 20th century science.

It is used in physics and most of chemistry to deal with problems about the atomic structure of matter. It is an extremely powerful mathematical tool and the whole basis of wave mechanics.

INTRODUCTION

In quantum mechanics, the **Schrödinger equation** is a mathematical equation, known as a partial differential equation, that describes the evolution over time of a physical system in which quantum effects, such a particle-wave duality, are significant. The equation is considered a central result in the study of quantum system and its derivation was a significant landmark in developing the theory of quantum mechanics. It is named after the Erwin Schrödinger published a paper with the equation in 1926. The publication of the equation had an influential effect on the physics community.

Schrödinger equation as part of a monument in front of Warsaw University's Centre of New Technologies.

In classical mechanics, Newton's second law (**F** = m**a**) is used to make a mathematical prediction as to what path a given system will take following a set of known initial conditions. In quantum mechanics, the analogue of Newton's law is Schrödinger's equation for a quantum system (usually atoms, molecules, and subatomic particles whether free, bound, or localised). It is not a simple algebraic equation, but in general a linear partial differential equation, describing the time-evolution of the system's wave function (also called a "state function").

The concept of a wavefunction is a fundamental postulate of quantum mechanics. Using these postulates, Schrödinger's equation can be derived from the fact that the time-evolution operator must be unitary and must therefore be generated by the exponential of a self-adjoint operator, which is the quantum Hamiltonian. This derivation is explained below.

In the Copenhagen interpretation of quantum mechanics, the wave function is the most complete description that can be given of a physical system. Solutions to Schrödinger's equation describe not only molecular, atomic, and subatomic systems, but also macroscopic systems, possibly even the whole universe. Schrödinger's equation is central to all applications of quantum mechanics including quantum field theory which combines special relativity with quantum mechanics. Theories of quantum gravity, such as string theory also do not modify Schrödinger's equation.

The Schrödinger equation is not the only way to make predictions in quantum mechanics—other formulations can be used, such as Werner Heisenberg's matrix mechanics, and Richard Feynman's path integral formulation.

EQUATION

Time-dependent equation

The form of the Schrödinger equation depends on the physical situation (see below for special cases). The most general form is the time-dependent Schrödinger equation, which gives a description of a system evolving with time:

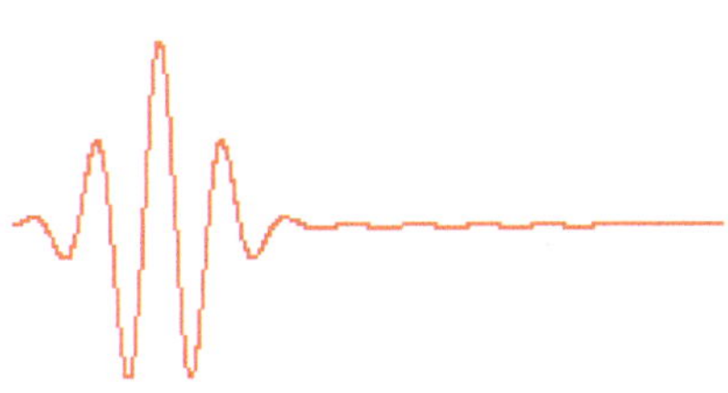

A wave function that satisfies the nonrelativistic Schrödinger equation with V = 0. In other words, this corresponds to a particle traveling freely through empty space. The real part of the wave function is plotted here.

Time-dependent Schrödinger equation *(general)*

$$i\hbar\frac{\partial}{\partial t}\Psi(\mathbf{r},t) = \hat{H}\Psi(\mathbf{r},t)$$

where i is the imaginary unit, ħ is the Planck constant divided by 2π, the symbol ∂/∂t indicates a partial derivative with respect to time t, Ψ (the Greek letter psi) is the wave function of the quantum system, **r** and tare the position vector and time respectively, and Ĥ is the Hamiltonian operator (which characterizes the total energy of any given wave function and takes different forms depending on the situation).

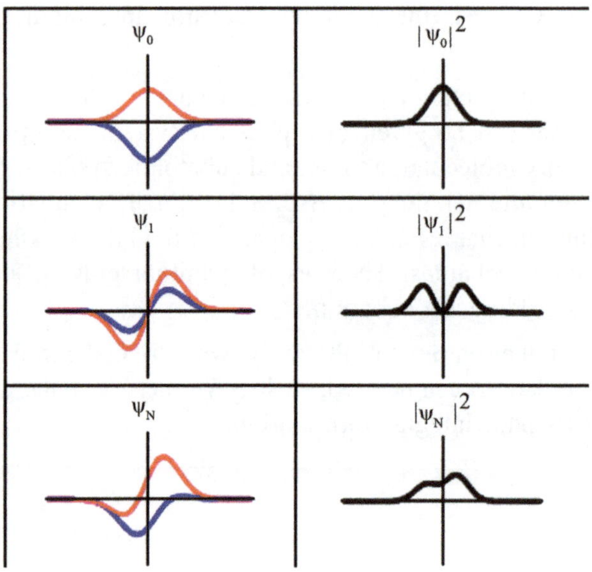

Each of these three rows is a wave function which satisfies the time-dependent Schrödinger equation for a harmonic oscillator. Left: The real part (blue) and imaginary part (red) of the wave function. Right: The probability distribution of finding the particle with this wave function at a given position. The top two rows are examples of **stationary states**, which correspond to standing waves. The bottom row is an example of a state which is not a stationary state. The right column illustrates why stationary states are called "stationary".

The most famous example is the nonrelativistic Schrödinger equation for a single particle moving in an electric field (but not a magnetic field; see the Pauli equation):

Time-dependent Schrödinger equation

(single nonrelativistic particle)

$$i\hbar\frac{\partial}{\partial t}\Psi(\mathbf{r},t) = \left[\frac{-\hbar^2}{2\mu}\nabla^2 + V(\mathbf{r},t)\right]\Psi(\mathbf{r},t)$$

where μ is the particle›s reduced mass", V is its potential energy, ∇^2 is the Laplacian (a differential operator), and Ψ is the wave function (more precisely, in this context, it is called the position-space wave function). In plain language, it means total energy equals kinetic energy plus potential energy", but the terms take unfamiliar forms for reasons explained below.

Given the particular differential operators involved, this is a linear partial differential equation. It is also a diffusion equation, but unlike the heat equation, this one is also a wave equation given the imaginary unit present in the transient term.

The term "Schrödinger equation" can refer to both the general equation (first box above), or the specific nonrelativistic version (second box above and variations thereof). The general equation is indeed quite general, used throughout quantum mechanics, for everything from the Dirac equation to quantum field theory, by plugging in various complicated expressions for the Hamiltonian. The specific nonrelativistic version is a simplified approximation to reality, which is quite accurate in many situations, but very inaccurate in others (see relativistic quantum mechanics and relativistic quantum field theory).

To apply the Schrödinger equation, the Hamiltonian operator is set up for the system, accounting for the kinetic and potential energy of the particles constituting the system, then inserted into the Schrödinger equation. The resulting partial differential equation is solved for the wave function, which contains information about the system.

Time-Independent Equation

The time-dependent Schrödinger equation described above predicts that wave functions can form standing waves, called stationary states (also called "orbitals", as in atomic orbitals or molecular orbitals). These states are important in their own right, and if the stationary states are classified and understood, then it becomes easier to solve the time-dependent Schrödinger equation for any state. Stationary states can also be described by a simpler form of the Schrödinger equation, the time-independent Schrödinger equation. (This is only used when the Hamiltonian itself is not dependent on time explicitly. However, even in this case the total wave function still has a time dependency.)

Time-independent Schrödinger equation (*general*)

$$\hat{H}\Psi = E\Psi$$

In words, the equation states:

When the Hamiltonian operator acts on a certain wave function Ψ, and the result is proportional to the same wave function Ψ, then Ψ is a stationary state, and the proportionality constant, E, is the energy of the state Ψ.

The time-independent Schrödinger equation is discussed further below. In linear algebra terminology, this equation is an eigenvalue equation.

As before, the most famous manifestation is the nonrelativistic Schrödinger equation for a single particle moving in an electric field (but not a magnetic field):

Time-independent Schrödinger equation (*single nonrelativistic particle*)

$$\left[\frac{-\hbar^2}{2\mu}\nabla^2 + V(\mathbf{r})\right]\Psi(\mathbf{r}) = E\Psi(\mathbf{r})$$

with definitions as above.

Derivation

In the modern understanding of quantum mechanics, Schrödinger's equation may be derived as follows.[6] If the wave-function at time t is given by $\Psi(t)$, then by the linearity of quantum mechanics the wave-function at time t' must be given by $\Psi(t') = U(t',t)\Psi(t)$, where $U(t',t)$ is a linear operator. Since time-evolution must preserve the norm of the wave-function, it follows that $U(t',t)$ must be a member of the unitary group of operators acting on wave-functions. We also know that when $t' = t$, we must have $U(t,t) = 1$. Therefore, expanding the operator $U(t',t)$ for t' close to t, we can write $U(t',t) = 1 - iH(t' - t)$ where H is a Hermitian operator. This follows from the fact that the Lie algebra corresponding to the unitary group comprises Hermitian operators. Taking the limit as the time-difference $t' - t$ becomes very small, we obtain Schrödinger's equation.

So far, H is only an abstract Hermitian operator. However using the correspondence principle it is possible to show that, in the classical limit, the expectation value of H is indeed the classical energy. The correspondence principle does not completely fix the form of the quantum Hamiltonian due to the uncertainty principle and therefore the precise form of the quantum Hamiltonian must be fixed empirically.

IMPLICATIONS

The Schrödinger equation and its solutions introduced a breakthrough in thinking about physics. Schrödinger's equation was the first of its type, and solutions led to consequences that were very unusual and unexpected for the time.

Total, kinetic, and Potential Energy

The overall form of the equation is not unusual or unexpected, as it uses the principle of the conservation of energy. The terms of the nonrelativistic Schrödinger equation can be interpreted as total energy of the system, equal to the system kinetic energy plus the system potential energy. In this respect, it is just the same as in classical physics.

Quantization

The Schrödinger equation predicts that if certain properties of a system are measured, the result may be quantized, meaning that only specific discrete values can occur. One example is energy quantization: the energy of an electron in an atom is always one of the quantized energy levels, a fact discovered via atomic spectroscopy. (Energy quantization is discussed below.) Another example is quantization of angular momentum. This was an assumption in the earlier Bohr model of the atom, but it is a prediction of the Schrödinger equation.

Another result of the Schrödinger equation is that not every measurement gives a quantized result in quantum mechanics. For example, position, momentum, time, and (in some situations) energy can have any value across a continuous range.

Measurement and Uncertainty

In classical mechanics, a particle has, at every moment, an exact position and an exact momentum. These values change deterministically as the particle moves according to Newton's laws. Under the Copenhagen interpretation of quantum mechanics, particles do not have exactly determined properties, and when they are measured, the result is randomly drawn from a probability distribution. The Schrödinger equation predicts what the probability distributions are, but fundamentally cannot predict the exact result of each measurement.

The Heisenberg uncertainty principle is the statement of the inherent measurement uncertainty in quantum mechanics. It states that the more precisely a particle's position is known, the less precisely its momentum is known, and vice versa.

The Schrödinger equation describes the (deterministic) evolution of the wave function of a particle. However, even if the wave function is known exactly, the result of a specific measurement on the wave function is uncertain.

Quantum Tunneling

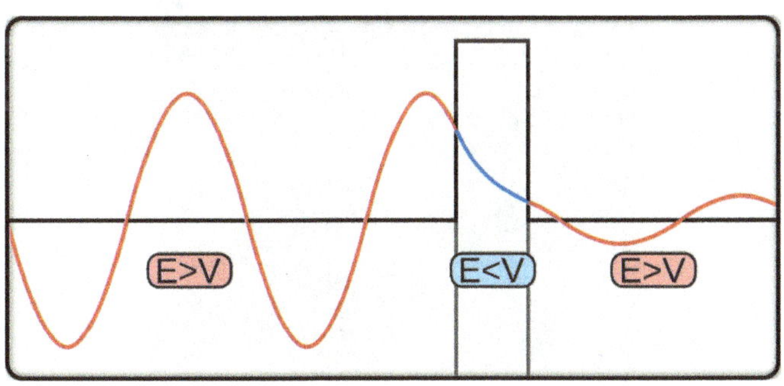

Quantum tunneling through a barrier. A particle coming from the left does not have enough energy to climb the barrier. However, it can sometimes "tunnel" to the other side.

In classical physics, when a ball is rolled slowly up a large hill, it will come to a stop and roll back, because it doesn't have enough energy to get over the top of the hill to the other side. However, the Schrödinger equation predicts that there is a small probability that the ball will get to the other side of the hill, even if it has too little energy to reach the top. This is called quantum tunneling. It is related to the distribution of energy: although the ball's assumed position seems to be on one side of the hill, there is a chance of finding it on the other side.

Particles as Waves

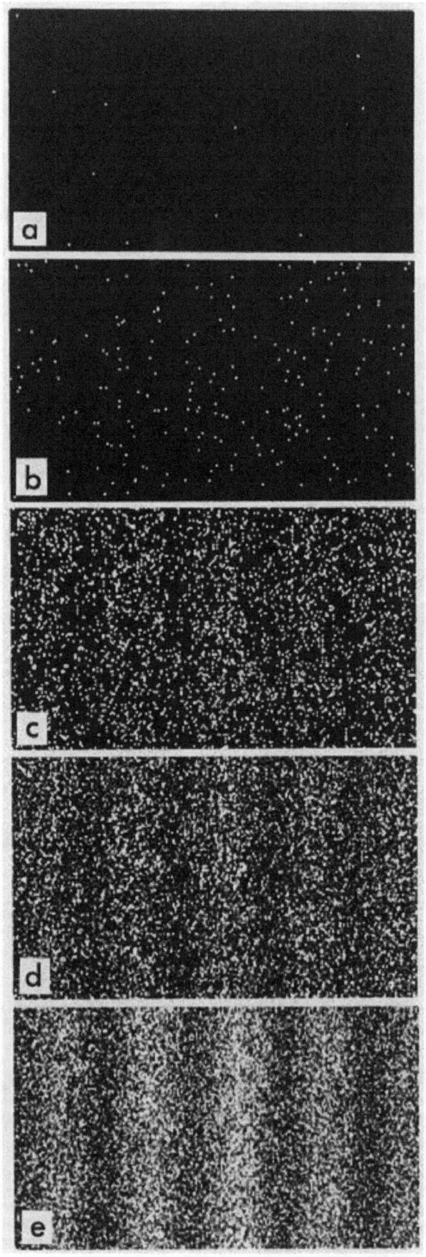

A double slit experiment showing the accumulation of electrons on a screen as time passes.

The nonrelativistic Schrödinger equation is a type of partial differential equation called a wave equation. Therefore, it is often said particles can exhibit behavior usually attributed to waves. In some modern interpretations this description is reversed – the quantum state, i.e. wave, is the only genuine physical reality, and under the appropriate conditions it can show features of particle-like behavior. However, Ballentine shows that such an interpretation has problems. Ballentine points out that whilst it is arguable to associate a physical wave with a single particle, there is still only one Schrödinger wave equation for many particles. He points out:

"If a physical wave field were associated with a particle, or if a particle were identified with a wave packet, then corresponding to N interacting particles there should be N interacting waves in ordinary three-dimensional space. But according to (4.6) that is not the case; instead there is one "wave" function in an abstract 3N-dimensional configuration space. The misinterpretation of psi as a physical wave in ordinary space is possible only because the most common applications of quantum mechanics are to one-particle states, for which configuration space and ordinary space are isomorphic."

Two-slit diffraction is a famous example of the strange behaviors that waves regularly display, that are not intuitively associated with particles. The overlapping waves from the two slits cancel each other out in some locations, and reinforce each other in other locations, causing a complex pattern to emerge. Intuitively, one would not expect this pattern from firing a single particle at the slits, because the particle should pass through one slit or the other, not a complex overlap of both.

However, since the Schrödinger equation is a wave equation, a single particle fired through a double-slit does show this same pattern (figure on right). Note: The experiment must be repeated many times for the complex pattern to emerge. Although this is counterintuitive, the prediction is correct; in particular, electron diffraction and neutron diffraction are well understood and widely used in science and engineering.

Related to diffraction, particles also display superposition and interference.

The superposition property allows the particle to be in a quantum superposition of two or more quantum states at the same time. However, it is noted that a "quantum state" in QM means the probability that a system will be, for example at a position x, not that the system will actually be at position x. It does not imply that the particle itself may be in two classical states at once. Indeed, QM is generally unable to assign values for properties prior to measurement at all.

Multiverse

In Dublin in 1952 Erwin Schrödinger gave a lecture in which at one point he jocularly warned his audience that what he was about to say might "seem lunatic". It was that, when his Nobel equations seem to be describing several different histories, they are "not alternatives but all really happen simultaneously". This is the earliest known reference to the multiverse.

INTERPRETATION OF THE WAVE FUNCTION

The Schrödinger equation provides a way to calculate the wave function of a system and how it changes dynamically in time. However, the Schrödinger equation does not directly say what, exactly, the wave function is. Interpretations of quantum mechanics address questions such as what the relation is between the wave function, the underlying reality, and the results of experimental measurements.

An important aspect is the relationship between the Schrödinger equation and wavefunction collapse. In the oldest Copenhagen interpretation, particles follow the Schrödinger equation except during wavefunction collapse, during which they behave entirely differently. The advent of quantum decoherence theory allowed alternative approaches (such as the Everett many-worlds interpretation and consistent histories), wherein the Schrödinger equation is always satisfied, and wavefunction collapse should be explained as a consequence of the Schrödinger equation.

HISTORICAL BACKGROUND AND DEVELOPMENT

Erwin Schrödinger.

Following Max Planck's quantization of light (see black body radiation), Albert Einstein interpreted Planck's quanta to be photons, particles of light, and proposed that the energy of a photon is proportional to its frequency, one of the first signs of wave–particle duality. Since energy and momentum are related in the same way as frequency and wavenumber in special relativity, it followed that the momentum p of a photon is inversely proportional to its wavelength λ, or proportional to its wavenumber k.

$$p = \frac{h}{\lambda} = \hbar k$$

where h is Planck's constant. Louis de Broglie hypothesized that this is true for all particles, even particles which have mass such as electrons. He showed that, assuming that the matter waves propagate along with their particle counterparts, electrons form standing waves, meaning that only certain discrete rotational frequencies about the nucleus of an atom are allowed. These quantized orbits correspond to discrete energy levels, and de Broglie reproduced the Bohr model formula for the energy levels. The Bohr model was based on the assumed quantization of angular momentum L according to:

$$L = n\frac{h}{2\pi} = n\hbar.$$

According to de Broglie the electron is described by a wave and a whole number of wavelengths must fit along the circumference of the electron's orbit:

$$n\lambda = 2\pi r.$$

This approach essentially confined the electron wave in one dimension, along a circular orbit of radius r.

In 1921, prior to de Broglie, Arthur C. Lunn at the University of Chicago had used the same argument based on the completion of the relativistic energy–momentum 4-vector to derive what we now call the de Broglie relation. Unlike de Broglie, Lunn went on to formulate the differential equation now known as the Schrödinger equation, and solve for its energy eigenvalues for the hydrogen atom. Unfortunately the paper was rejected by the Physical Review, as recounted by Kamen.

Following up on de Broglie's ideas, physicist Peter Debye made an offhand comment that if particles behaved as waves, they should satisfy some sort of wave equation. Inspired by Debye's remark, Schrödinger decided to find a proper 3-dimensional wave equation for the electron. He was guided by William R. Hamilton's analogy between mechanics and optics, encoded in the observation that the zero-wavelength limit of optics resembles a mechanical system—the trajectories of light rays become sharp tracks that obey Fermat's principle, an analog of the principle of least action. A modern version of his reasoning is reproduced below. The equation he found is:

$$i\hbar\frac{\partial}{\partial t}\Psi(\mathbf{r}, t) = -\frac{\hbar^2}{2m}\nabla^2\Psi(\mathbf{r}, t) + V(\mathbf{r})\Psi(\mathbf{r}, t).$$

However, by that time, Arnold Sommerfeld had refined the Bohr model with relativistic corrections. Schrödinger used the relativistic energy momentum relation to find what is now known as the Klein–Gordon equation in a Coulomb potential (in natural units):

$$\left(E + \frac{e^2}{r}\right)^2 \psi(x) = -\nabla^2\psi(x) + m^2\psi(x)$$

He found the standing waves of this relativistic equation, but the relativistic corrections disagreed with Sommerfeld's formula. Discouraged, he put away his calculations and secluded himself in an isolated mountain cabin in December 1925.

While at the cabin, Schrödinger decided that his earlier nonrelativistic calculations were novel enough to publish, and decided to leave off the problem of relativistic corrections for the future. Despite the difficulties in solving the differential equation for hydrogen (he had sought help from his friend the mathematician Hermann Weyl) Schrödinger showed that his nonrelativistic version of the wave equation produced the correct spectral energies of hydrogen in a paper published in 1926. In the equation, Schrödinger computed the hydrogen spectral series by treating a hydrogen atom's electron as a wave $\Psi(x, t)$, moving in a potential well V, created by the proton. This computation accurately reproduced the energy levels of the Bohr model. In a paper, Schrödinger himself explained this equation as follows:

This 1926 paper was enthusiastically endorsed by Einstein, who saw the matter-waves as an intuitive depiction of nature, as opposed to Heisenberg's matrix mechanics, which he considered overly formal.

The Schrödinger equation details the behavior of Ψ but says nothing of its nature. Schrödinger tried to interpret it as a charge density in his fourth paper, but he was unsuccessful.[:219] In 1926, just a few days after Schrödinger's fourth and final paper was published, Max Born successfully interpreted Ψ as the probability amplitude, whose absolute square is equal to probability density. Schrödinger, though, always opposed a statistical or probabilistic approach, with its associated discontinuities—much like Einstein, who believed that quantum mechanics was a statistical approximation to an underlying deterministic theory—and never reconciled with the Copenhagen interpretation.

Louis de Broglie in his later years proposed a real valued wave function connected to the complex wave function by a proportionality constant and developed the De Broglie–Bohm theory.

THE WAVE EQUATION FOR PARTICLES

The Schrödinger equation is a diffusion equation, the solutions are functions which describe wave-like motions. Wave equations in physics can normally be derived from other physical laws – the wave equation for mechanical vibrations on strings and in matter can be derived from Newton's laws – where the wave function represents the displacement of matter, and electromagnetic waves from Maxwell's equations, where the wave functions are electric and magnetic fields. The basis for Schrödinger's equation, on the other hand, is the energy of the system and a separate postulate of quantum mechanics: the wave function is a description of the system. The Schrödinger equation is therefore a new concept in itself; as Feynman put it:

The foundation of the equation is structured to be a linear differential equation based on classical energy conservation, and consistent with the De Broglie relations. The solution is the wave function ψ, which contains all the information that can be known about the system. In the Copenhagen interpretation, the modulus of ψ is related to the probability the particles are in some spatial configuration at some instant of time. Solving the equation for ψ can be used to predict how the particles will behave under the influence of the specified potential and with each other.

The Schrödinger equation was developed principally from the De Broglie hypothesis, a wave equation that would describe particles, and can be constructed as shown informally in the following sections. For a more rigorous description of Schrödinger's equation, see also Resnick et al.

Consistency with Energy Conservation

The total energy E of a particle is the sum of kinetic energy T and potential energy V, this sum is also the frequent expression for the Hamiltonian Hin classical mechanics:

$$E = T + V = H$$

Explicitly, for a particle in one dimension with position x, mass m and momentum p, and potential energy V which generally varies with position and time t:

$$E = \frac{p^2}{2m} + V(x,t) = H.$$

For three dimensions, the position vector **r** and momentum vector **p** must be used:

$$E = \frac{\mathbf{p}\cdot\mathbf{p}}{2m} + V(\mathbf{r},t) = H$$

This formalism can be extended to any fixed number of particles: the total energy of the system is then the total kinetic energies of the particles, plus the total potential energy, again the Hamiltonian. However, there can be interactions between the particles (an N-body problem), so the potential energy V can change as the spatial configuration of particles changes, and possibly with time. The potential energy, in general, is not the sum of the separate potential energies for each particle, it is a function of all the spatial positions of the particles. Explicitly:

$$E = \sum_{n=1}^{N} \frac{\mathbf{p}_n\cdot\mathbf{p}_n}{2m_n} + V(\mathbf{r}_1,\mathbf{r}_2\cdots\mathbf{r}_N,t) = H$$

Linearity

The simplest wavefunction is a plane wave of the form:

$$\Psi(\mathbf{r},t) = Ae^{i(\mathbf{k}\cdot\mathbf{r}-\omega t)}$$

where the A is the amplitude, **k** the wavevector, and ω the angular frequency, of the plane wave. In general, physical situations are not purely described by plane waves, so for generality the superposition principle is required; any wave can be made by superposition of sinusoidal plane waves. So if the equation is linear, a linear combination of plane waves is also an allowed solution. Hence a necessary and separate requirement is that the Schrödinger equation is a linear differential equation.

For discrete **k** the sum is a superposition of plane waves:

$$\Psi(\mathbf{r},t)=\sum_{n=1}^{\infty}A_n e^{i(\mathbf{k}_n\cdot\mathbf{r}-\omega_n t)}$$

for some real amplitude coefficients A_n, and for continuous **k** the sum becomes an integral, the Fourier transform of a momentum space wavefunction:

$$\Psi(\mathbf{r},t)=\frac{1}{(\sqrt{2\pi})^3}\int\Phi(\mathbf{k})e^{i(\mathbf{k}\cdot\mathbf{r}-\omega t)}d^3\mathbf{k}$$

where $d^3\mathbf{k} = dk_x dk_y dk_z$ is the differential volume element in **k**-space, and the integrals are taken over all **k**-space. The momentum wavefunction $\Phi(\mathbf{k})$ arises in the integrand since the position and momentum space wavefunctions are Fourier transforms of each other.

Consistency with the De Broglie Relations

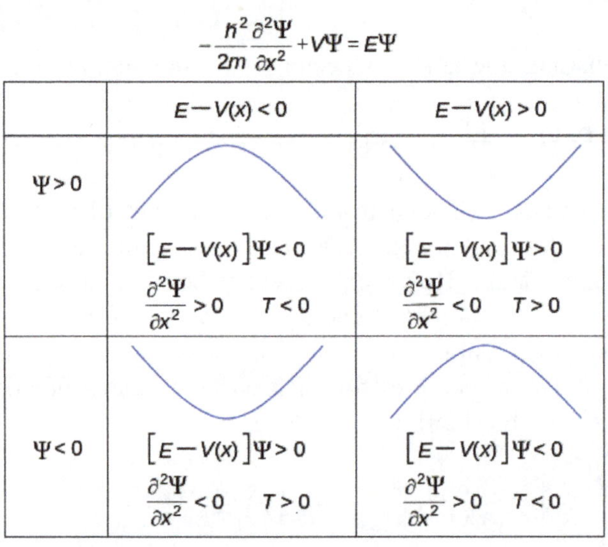

Diagrammatic summary of the quantities related to the wavefunction, as used in De broglie's hypothesis and development of the Schrödinger equation.

Einstein's light quanta hypothesis (1905) states that the energy E of a photon is proportional to the frequency ν (or angular frequency, ω = 2πν) of the corresponding quantum wavepacket of light:

$$E=h\nu=\hbar\omega$$

Likewise De Broglie's hypothesis (1924) states that any particle can be associated with a wave, and that the momentum p of the particle is inversely proportional to the wavelength λ of such a wave (or proportional to the wavenumber, k = 2π/λ), in one dimension, by:

$$p = \frac{h}{\lambda} = \hbar k \,,$$

while in three dimensions, wavelength λ is related to the magnitude of the wavevector **k**:

$$\mathbf{p} = \hbar \mathbf{k} \,, \quad |\mathbf{k}| = \frac{2\pi}{\lambda} \,.$$

The Planck–Einstein and de Broglie relations illuminate the deep connections between energy with time, and space with momentum, and express wave–particle duality. In practice, natural units comprising ħ = 1 are used, as the De Broglie equations reduce to identities: allowing momentum, wavenumber, energy and frequency to be used interchangeably, to prevent duplication of quantities, and reduce the number of dimensions of related quantities. For familiarity SI units are still used in this article.

Schrödinger's insight, late in 1925, was to express the phase of a plane wave as a complex phase factor using these relations:

$$\Psi = Ae^{i(\mathbf{k}\cdot\mathbf{r}-\omega t)} = Ae^{i(\mathbf{p}\cdot\mathbf{r}-Et)/\hbar}$$

and to realize that the first order partial derivatives were:

with respect to space:

$$\nabla\Psi = \frac{i}{\hbar}\mathbf{p}Ae^{i(\mathbf{p}\cdot\mathbf{r}-Et)/\hbar} = \frac{i}{\hbar}\mathbf{p}\Psi$$

with respect to time:

$$\frac{\partial\Psi}{\partial t} = -\frac{iE}{\hbar}Ae^{i(\mathbf{p}\cdot\mathbf{r}-Et)/\hbar} = -\frac{iE}{\hbar}\Psi$$

Another postulate of quantum mechanics is that all observables are represented by linear Hermitian operators which act on the wavefunction, and the eigenvalues of the operator are the values the observable takes. The previous derivatives are consistent with the energy operator, corresponding to the time derivative,

$$\hat{E}\Psi = i\hbar\frac{\partial}{\partial t}\Psi = E\Psi$$

where E are the energy eigenvalues, and the momentum operator, corresponding to the spatial derivatives (the gradient ∇),

$$\hat{\mathbf{p}}\Psi = -i\hbar\nabla\Psi = \mathbf{p}\Psi$$

where **p** is a vector of the momentum eigenvalues. In the above, the "hats" (^) indicate these observables are operators, not simply ordinary numbers or vectors. The energy and momentum operators are differential operators, while the potential energy function V is just a multiplicative factor.

Substituting the energy and momentum operators into the classical energy conservation equation obtains the operator:

$$E = \frac{\mathbf{p}\cdot\mathbf{p}}{2m} + V \quad \rightarrow \quad \hat{E} = \frac{\hat{\mathbf{p}}\cdot\hat{\mathbf{p}}}{2m} + V$$

so in terms of derivatives with respect to time and space, acting this operator on the wavefunction Ψ immediately led Schrödinger to his equation:

$$i\hbar\frac{\partial\Psi}{\partial t} = -\frac{\hbar^2}{2m}\nabla^2\Psi + V\Psi$$

Wave–particle duality can be assessed from these equations as follows. The kinetic energy T is related to the square of momentum **p**. As the particle's momentum increases, the kinetic energy increases more rapidly, but since the wavenumber |**k**| increases the wavelength λ decreases. In terms of ordinary scalar and vector quantities (not operators):

$$\mathbf{p}\cdot\mathbf{p} \propto \mathbf{k}\cdot\mathbf{k} \propto T \propto \frac{1}{\lambda^2}$$

The kinetic energy is also proportional to the second spatial derivatives, so it is also proportional to the magnitude of the curvature of the wave, in terms of operators:

$$\hat{T}\Psi = \frac{-\hbar^2}{2m}\nabla\cdot\nabla\Psi \propto \nabla^2\Psi.$$

As the curvature increases, the amplitude of the wave alternates between positive and negative more rapidly, and also shortens the wavelength. So the inverse relation between momentum and wavelength is consistent with the energy the particle has, and so the energy of the particle has a connection to a wave, all in the same mathematical formulation.

Wave and Particle Motion

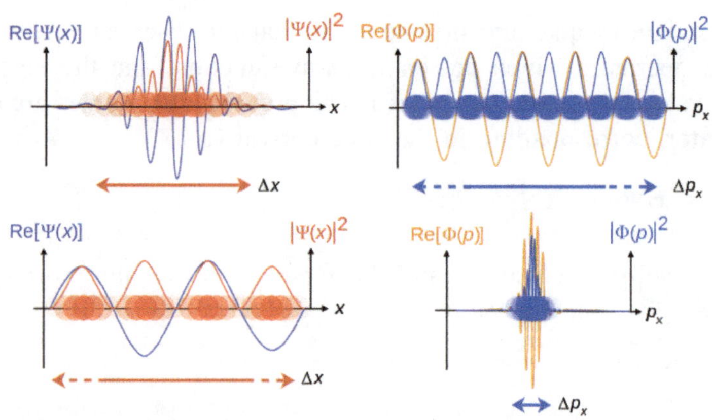

Increasing levels of wavepacket localization, meaning the particle has a more localized position.

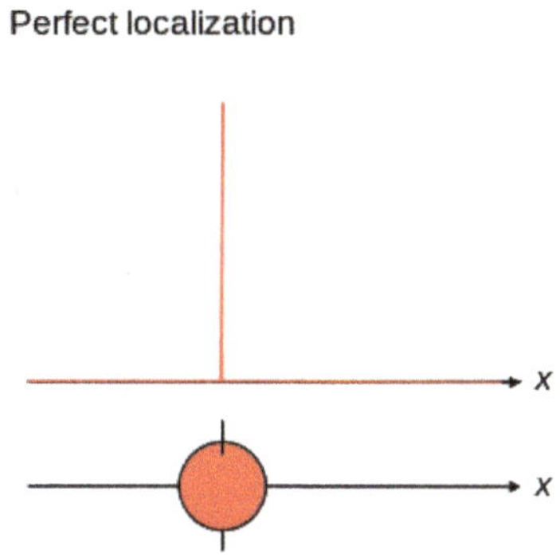

In the limit ħ → 0, the particle›s position and momentum become known exactly. This is equivalent to the classical particle.

Schrödinger required that a wave packet solution near position **r** with wavevector near **k** will move along the trajectory determined by classical mechanics for times short enough for the spread in **k**(and hence in velocity) not to substantially increase the spread in **r**. Since, for a given spread in **k**, the spread in velocity is proportional to Planck's constant ħ, it is sometimes said that in the limit as ħapproaches zero, the equations of classical mechanics are restored from quantum mechanics.Great care is required in how that limit is taken, and in what cases.

The limiting short-wavelength is equivalent to ħtending to zero because this is limiting case of increasing the wave packet localization to the definite position of the particle (see images right). Using the Heisenberg uncertainty principle for position and momentum, the products of uncertainty in position and momentum become zero as ħ → 0:

$$\sigma(x)\sigma(p_x) \geqslant \frac{\hbar}{2} \quad \rightarrow \quad \sigma(x)\sigma(p_x) \geqslant 0$$

where σ denotes the (root mean square) measurement uncertainty in x and p_x (and similarly for the y and z directions) which implies the position and momentum can only be known to arbitrary precision in this limit.

The Schrödinger equation in its general form

$$i\hbar\frac{\partial}{\partial t}\Psi(\mathbf{r},t) = \hat{H}\Psi(\mathbf{r},t)$$

is closely related to the Hamilton–Jacobi equation (HJE)

$$\frac{\partial}{\partial t}S(q_i,t) = H\left(q_i, \frac{\partial S}{\partial q_i}, t\right)$$

where S is action and H is the Hamiltonian function (not operator). Here the generalized coordinates q_i for i = 1, 2, 3 (used in the context of the HJE) can be set to the position in Cartesian coordinates as **r** = (q_1, q_2, q_3) = (x, y, z).

Substituting

$$\Psi = \sqrt{\rho(\mathbf{r}, t)} e^{iS(\mathbf{r},t)/\hbar}$$

where ρ is the probability density, into the Schrödinger equation and then taking the limit $\hbar \to 0$ in the resulting equation, yields the Hamilton–Jacobi equation.

The implications are:

- The motion of a particle, described by a (short-wavelength) wave packet solution to the Schrödinger equation, is also described by the Hamilton–Jacobi equation of motion.
- The Schrödinger equation includes the wavefunction, so its wave packet solution implies the position of a (quantum) particle is fuzzily spread out in wave fronts. On the contrary, the Hamilton–Jacobi equation applies to a (classical) particle of definite position and momentum, instead the position and momentum at all times (the trajectory) are deterministic and can be simultaneously known.

NONRELATIVISTIC QUANTUM MECHANICS

The quantum mechanics of particles without accounting for the effects of special relativity, for example particles propagating at speeds much less than light, is known as **nonrelativistic quantum mechanics**. Following are several forms of Schrödinger's equation in this context for different situations: time independence and dependence, one and three spatial dimensions, and one and N particles.

In actuality, the particles constituting the system do not have the numerical labels used in theory. The language of mathematics forces us to label the positions of particles one way or another, otherwise there would be confusion between symbols representing which variables are for which particle.

Time Independent

If the Hamiltonian is not an explicit function of time, the equation is separable into a product of spatial and temporal parts. In general, the wavefunction takes the form:

$$\Psi(\text{space coords}, t) = \psi(\text{space coords})\tau(t)$$

where ψ(space coords) is a function of all the spatial coordinate(s) of the particle(s) constituting the system only, and τ(t) is a function of time only.

Substituting for ψ into the Schrödinger equation for the relevant number of particles in the relevant number of dimensions, solving by separation of variables implies the general solution of the time-dependent equation has the form:

$$\Psi(\text{space coords}, t) = \psi(\text{space coords})e^{-iEt/\hbar}.$$

Since the time dependent phase factor is always the same, only the spatial part needs to be solved for in time independent problems. Additionally, the energy operator $\hat{E} = i\hbar\partial/\partial t$ can always be replaced by the energy eigenvalue E, thus the time independent Schrödinger equation is an eigenvalue equation for the Hamiltonian operator:[143ff]

$$\hat{H}\psi = E\psi$$

This is true for any number of particles in any number of dimensions (in a time independent potential). This case describes the standing wave solutions of the time-dependent equation, which are the states with definite energy (instead of a probability distribution of different energies). In physics, these standing waves are called "stationary states" or "energy eigenstates"; in chemistry they are called "atomic orbitals" or "molecular orbitals". Superpositions of energy eigenstates change their properties according to the relative phases between the energy levels.

The energy eigenvalues from this equation form a discrete spectrum of values, so mathematically energy must be quantized. More specifically, the energy eigenstates form a basis – any wavefunction may be written as a sum over the discrete energy states or an integral over continuous energy states, or more generally as an integral over a measure. This is the spectral theorem in mathematics, and in a finite state space it is just a statement of the completeness of the eigenvectors of a Hermitian matrix.

One-Dimensional Examples

For a particle in one dimension, the Hamiltonian is:

$$\hat{H} = \frac{\hat{p}^2}{2m} + V(x), \quad \hat{p} = -i\hbar\frac{d}{dx}$$

and substituting this into the general Schrödinger equation gives:

$$-\frac{\hbar^2}{2m}\frac{d^2}{dx^2}\psi(x) + V(x)\psi(x) = E\psi(x)$$

This is the only case the Schrödinger equation is an ordinary differential equation, rather than a partial differential equation. The general solutions are always of the form:

$$\Psi(x,t) = \psi(x)e^{-iEt/\hbar}.$$

For N particles in one dimension, the Hamiltonian is:

$$\hat{H} = \sum_{n=1}^{N}\frac{\hat{p}_n^2}{2m_n} + V(x_1, x_2, \cdots x_N), \quad \hat{p}_n = -i\hbar\frac{\partial}{\partial x_n}$$

where the position of particle n is x_n. The corresponding Schrödinger equation is:

$$-\frac{\hbar^2}{2}\sum_{n=1}^{N}\frac{1}{m_n}\frac{\partial^2}{\partial x_n^2}\psi(x_1, x_2, \cdots x_N) + V(x_1, x_2, \cdots x_N)\psi(x_1, x_2, \cdots x_N) = E\psi(x_1, x_2, \cdots x_N).$$

so the general solutions have the form:

$$\Psi(x_1, x_2, \cdots x_N, t) = e^{-iEt/\hbar}\psi(x_1, x_2 \cdots x_N)$$

For non-interacting distinguishable particles, the potential of the system only influences each particle separately, so the total potential energy is the sum of potential energies for each particle:

$$V(x_1, x_2, \cdots x_N) = \sum_{n=1}^{N} V(x_n).$$

and the wavefunction can be written as a product of the wavefunctions for each particle:

$$\Psi(x_1, x_2, \cdots x_N, t) = e^{-iEt/\hbar} \prod_{n=1}^{N} \psi(x_n),$$

For non-interacting identical particles, the potential is still a sum, but wavefunction is a bit more complicated – it is a sum over the permutations of products of the separate wavefunctions to account for particle exchange. In general for interacting particles, the above decompositions are not possible.

Free Particle

For no potential, V = 0, so the particle is free and the equation reads:[:151ff]

$$-E\psi = \frac{\hbar^2}{2m}\frac{d^2\psi}{dx^2}$$

which has oscillatory solutions for E > 0 (the C_n are arbitrary constants):

$$\psi_E(x) = C_1 e^{i\sqrt{2mE/\hbar^2}\,x} + C_2 e^{-i\sqrt{2mE/\hbar^2}\,x}$$

and exponential solutions for E < 0

$$\psi_{-|E|}(x) = C_1 e^{\sqrt{2m|E|/\hbar^2}\,x} + C_2 e^{-\sqrt{2m|E|/\hbar^2}\,x}.$$

The exponentially growing solutions have an infinite norm, and are not physical. They are not allowed in a finite volume with periodic or fixed boundary conditions.

See also free particle and wavepacket for more discussion on the free particle.

Constant Potential

For a constant potential, $V = V_0$, the solution is oscillatory for $E > V_0$ and exponential for $E < V_0$, corresponding to energies that are allowed or disallowed in classical mechanics. Oscillatory solutions have a classically allowed energy and correspond to actual classical motions, while the exponential solutions have a disallowed energy and describe a small amount of quantum bleeding into the classically disallowed region, due to quantum tunneling. If the potential V_0 grows to infinity, the motion is classically confined to a finite region. Viewed far enough away, every solution is reduced to an exponential; the condition that the exponential is decreasing restricts the energy levels to a discrete set, called the allowed energies.

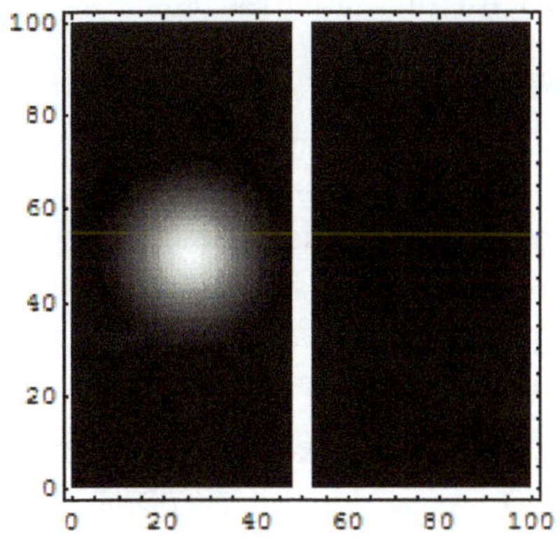

Harmonic Oscillator

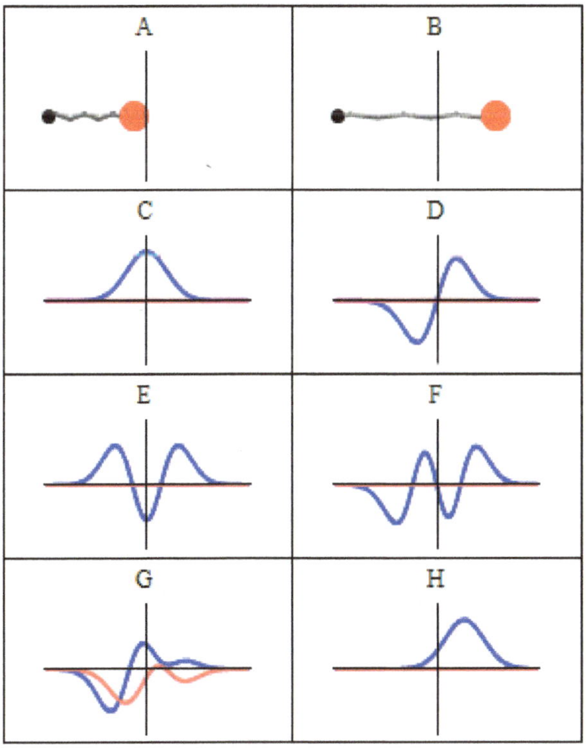

A harmonic oscillator in classical mechanics (A–B) and quantum mechanics (C–H). In (A–B), a ball, attached to a spring, oscillates back and forth. (C–H) are six solutions to the Schrödinger Equation for this situation. The horizontal axis is position, the vertical axis is the real part (blue) or imaginary part (red) of the wavefunction. Stationary states, or energy eigenstates, which are solutions to the time-independent Schrödinger equation, are shown in C, D, E, F, but not G or H.

The Schrödinger equation for this situation is

$$E\psi = -\frac{\hbar^2}{2m}\frac{d^2}{dx^2}\psi + \frac{1}{2}m\omega^2 x^2\psi$$

It is a notable quantum system to solve for; since the solutions are exact (but complicated – in terms of Hermite polynomials), and it can describe or at least approximate a wide variety of other systems, including vibrating atoms, molecules, and atoms or ions in lattices, and approximating other potentials near equilibrium points. It is also the basis of perturbation methods in quantum mechanics.

There is a family of solutions – in the position basis they are

$$\psi_n(x) = \sqrt{\frac{1}{2^n\, n!}} \cdot \left(\frac{m\omega}{\pi\hbar}\right)^{1/4} \cdot e^{-\frac{m\omega x^2}{2\hbar}} \cdot H_n\left(\sqrt{\frac{m\omega}{\hbar}}x\right)$$

where n = 0,1,2,..., and the functions H_n are the Hermite polynomials.

Three-dimensional Examples

The extension from one dimension to three dimensions is straightforward, all position and momentum operators are replaced by their three-dimensional expressions and the partial derivative with respect to space is replaced by the gradient operator.

The Hamiltonian for one particle in three dimensions is:

$$\hat{H} = \frac{\hat{\mathbf{p}}\cdot\hat{\mathbf{p}}}{2m} + V(\mathbf{r}), \quad \hat{\mathbf{p}} = -i\hbar\nabla$$

generating the equation:

$$-\frac{\hbar^2}{2m}\nabla^2\psi(\mathbf{r}) + V(\mathbf{r})\psi(\mathbf{r}) = E\psi(\mathbf{r})$$

with stationary state solutions of the form:

$$\Psi(\mathbf{r},t) = \psi(\mathbf{r})e^{-iEt/\hbar}$$

where the position of the particle is **r**. Two useful coordinate systems for solving the Schrödinger equation are Cartesian coordinates so that **r** = (x, y, z) and spherical polar coordinates so that **r** = (r, θ, φ), although other orthogonal coordinates are useful for solving the equation for systems with certain geometric symmetries.

For N particles in three dimensions, the Hamiltonian is:

$$\hat{H} = \sum_{n=1}^{N}\frac{\hat{\mathbf{p}}_n\cdot\hat{\mathbf{p}}_n}{2m_n} + V(\mathbf{r}_1,\mathbf{r}_2,\cdots\mathbf{r}_N), \quad \hat{\mathbf{p}}_n = -i\hbar\nabla_n$$

where the position of particle n is $\mathbf{r}_n$ and the gradient operators are partial derivatives with respect to the particle's position coordinates. In Cartesian coordinates, for particle n, the position vector is $\mathbf{r}_n = (x_n, y_n, z_n)$ while the gradient and Laplacian operator are respectively:

$$\nabla_n = \mathbf{e}_x\frac{\partial}{\partial x_n} + \mathbf{e}_y\frac{\partial}{\partial y_n} + \mathbf{e}_z\frac{\partial}{\partial z_n}, \quad \nabla_n^2 = \nabla_n\cdot\nabla_n = \frac{\partial^2}{\partial {x_n}^2} + \frac{\partial^2}{\partial {y_n}^2} + \frac{\partial^2}{\partial {z_n}^2}$$

The Schrödinger equation is:

$$-\frac{\hbar^2}{2}\sum_{n=1}^{N}\frac{1}{m_n}\nabla_n^2\Psi(\mathbf{r}_1,\mathbf{r}_2,\cdots\mathbf{r}_N)+V(\mathbf{r}_1,\mathbf{r}_2,\cdots\mathbf{r}_N)\Psi(\mathbf{r}_1,\mathbf{r}_2,\cdots\mathbf{r}_N)=E\Psi(\mathbf{r}_1,\mathbf{r}_2,\cdots\mathbf{r}_N)$$

with stationary state solutions:

$$\Psi(\mathbf{r}_1,\mathbf{r}_2\cdots\mathbf{r}_N,t)=e^{-iEt/\hbar}\psi(\mathbf{r}_1,\mathbf{r}_2\cdots\mathbf{r}_N)$$

Again, for non-interacting distinguishable particles the potential is the sum of particle potentials

$$V(\mathbf{r}_1,\mathbf{r}_2,\cdots\mathbf{r}_N)=\sum_{n=1}^{N}V(\mathbf{r}_n)$$

and the wavefunction is a product of the particle wavefuntions

$$\Psi(\mathbf{r}_1,\mathbf{r}_2\cdots\mathbf{r}_N,t)=e^{-iEt/\hbar}\prod_{n=1}^{N}\psi(\mathbf{r}_n)$$

For non-interacting identical particles, the potential is a sum but the wavefunction is a sum over permutations of products. The previous two equations do not apply to interacting particles.

Following are examples where exact solutions are known. See the main articles for further details.

Hydrogen Atom

This form of the Schrödinger equation can be applied to the hydrogen atom:

$$E\psi=-\frac{\hbar^2}{2\mu}\nabla^2\psi-\frac{e^2}{4\pi\varepsilon_0 r}\psi$$

where e is the electron charge, **r** is the position of the electron ($r = |\mathbf{r}|$ is the magnitude of the position), the potential term is due to the Coulomb interaction, wherein ε_0 is the electric constant (permittivity of free space) and

$$\mu=\frac{m_e m_p}{m_e+m_p}$$

is the 2-body reduced mass of the hydrogen nucleus (just a proton) of mass m_p and the electron of mass m_e. The negative sign arises in the potential term since the proton and electron are oppositely charged. The reduced mass in place of the electron mass is used since the electron and proton together orbit each other about a common centre of mass, and constitute a two-body problem to solve. The motion of the electron is of principle interest here, so the equivalent one-body problem is the motion of the electron using the reduced mass.

The wavefunction for hydrogen is a function of the electron's coordinates, and in fact can be separated into functions of each coordinate.Usually this is done in spherical polar coordinates:

$$\psi(r,\theta,\phi) = R(r)Y_\ell^m(\theta,\phi) = R(r)\Theta(\theta)\Phi(\phi)$$

where R are radial functions and Ymℓ(θ, φ) are spherical harmonics of degree ℓ and order m. This is the only atom for which the Schrödinger equation has been solved for exactly. Multi-electron atoms require approximative methods. The family of solutions are:

$$\psi_{n\ell m}(r,\theta,\phi) = \sqrt{\left(\frac{2}{na_0}\right)^3 \frac{(n-\ell-1)!}{2n[(n+\ell)!]}} e^{-r/na_0} \left(\frac{2r}{na_0}\right)^\ell L_{n-\ell-1}^{2\ell+1}\left(\frac{2r}{na_0}\right) \cdot Y_\ell^m(\theta,\phi)$$

where:

- $a_0 = \frac{4\pi\varepsilon_0\hbar^2}{m_e e^2}$ is the Bohr radius,
- $L_{n-\ell-1}^{2\ell+1}(\cdots)$ are the generalized Laguerre polynomials of degree $n-\ell-1$.
- n, ℓ, m are the principal, azimuthal, and magnetic quantum numbers respectively: which take the values:

$$n = 1, 2, 3, \ldots$$
$$\ell = 0, 1, 2, \ldots, n-1$$
$$m = -\ell, \ldots, \ell$$

NB: generalized Laguerre polynomials are defined differently by different authors—see main article on them and the hydrogen atom.

Two-electron Atoms or Ions

The equation for any two-electron system, such as the neutral helium atom (He, Z = 2), the negative hydrogen ion (H^-, Z = 1), or the positive lithium ion (Li^+, Z = 3) is:

$$E\psi = -\hbar^2\left[\frac{1}{2\mu}(\nabla_1^2 + \nabla_2^2) + \frac{1}{M}\nabla_1 \cdot \nabla_2\right]\psi + \frac{e^2}{4\pi\varepsilon_0}\left[\frac{1}{r_{12}} - Z\left(\frac{1}{r_1} + \frac{1}{r_2}\right)\right]\psi$$

where $\mathbf{r}_1$ is the position of one electron ($r_1 = |\mathbf{r}_1|$ is its magnitude), $\mathbf{r}_2$ is the position of the other electron ($r_2 = |\mathbf{r}_2|$ is the magnitude), $r_{12} = |\mathbf{r}_{12}|$ is the magnitude of the separation between them given by

$$|\mathbf{r}_{12}| = |\mathbf{r}_2 - \mathbf{r}_1|$$

μ is again the two-body reduced mass of an electron with respect to the nucleus of mass M, so this time

$$\mu = \frac{m_e M}{m_e + M}$$

and Z is the atomic number for the element (not a quantum number).

The cross-term of two laplacians

$$\frac{1}{M}\nabla_1 \cdot \nabla_2$$

is known as the mass polarization term, which arises due to the motion of atomic nuclei. The wavefunction is a function of the two electron's positions:

$$\psi = \psi(\mathbf{r}_1, \mathbf{r}_2).$$

There is no closed form solution for this equation.

Time Dependent

This is the equation of motion for the quantum state. In the most general form, it is written:[:143ff]

$$i\hbar\frac{\partial}{\partial t}\Psi = \hat{H}\Psi,$$

and the solution, the wavefunction, is a function of all the particle coordinates of the system and time. Following are specific cases.

For one particle in one dimension, the Hamiltonian

$$\hat{H} = \frac{\hat{p}^2}{2m} + V(x,t)\,, \quad \hat{p} = -i\hbar\frac{\partial}{\partial x}$$

generates the equation:

$$i\hbar\frac{\partial}{\partial t}\Psi(x,t) = -\frac{\hbar^2}{2m}\frac{\partial^2}{\partial x^2}\Psi(x,t) + V(x,t)\Psi(x,t)$$

For N particles in one dimension, the Hamiltonian is:

$$\hat{H} = \sum_{n=1}^{N}\frac{\hat{p}_n^2}{2m_n} + V(x_1,x_2,\cdots x_N,t)\,, \quad \hat{p}_n = -i\hbar\frac{\partial}{\partial x_n}$$

where the position of particle n is x_n, generating the equation:

$$i\hbar\frac{\partial}{\partial t}\Psi(x_1,x_2\cdots x_N,t) = -\frac{\hbar^2}{2}\sum_{n=1}^{N}\frac{1}{m_n}\frac{\partial^2}{\partial x_n^2}\Psi(x_1,x_2\cdots x_N,t) + V(x_1,x_2\cdots x_N,t)\Psi(x_1,x_2\cdots x_N,t).$$

For one particle in three dimensions, the Hamiltonian is:

$$\hat{H} = \frac{\hat{\mathbf{p}}\cdot\hat{\mathbf{p}}}{2m} + V(\mathbf{r},t)\,, \quad \hat{\mathbf{p}} = -i\hbar\nabla$$

generating the equation:

$$i\hbar\frac{\partial}{\partial t}\Psi(\mathbf{r},t) = -\frac{\hbar^2}{2m}\nabla^2\Psi(\mathbf{r},t) + V(\mathbf{r},t)\Psi(\mathbf{r},t)$$

For N particles in three dimensions, the Hamiltonian is:

$$\hat{H} = \sum_{n=1}^{N}\frac{\hat{\mathbf{p}}_n\cdot\hat{\mathbf{p}}_n}{2m_n} + V(\mathbf{r}_1,\mathbf{r}_2,\cdots\mathbf{r}_N,t)\,, \quad \hat{\mathbf{p}}_n = -i\hbar\nabla_n$$

where the position of particle n is $\mathbf{r}_n$, generating the equation:[4]:141

$$i\hbar\frac{\partial}{\partial t}\Psi(\mathbf{r}_1,\mathbf{r}_2,\cdots\mathbf{r}_N,t) = -\frac{\hbar^2}{2}\sum_{n=1}^{N}\frac{1}{m_n}\nabla_n^2\Psi(\mathbf{r}_1,\mathbf{r}_2,\cdots\mathbf{r}_N,t) + V(\mathbf{r}_1,\mathbf{r}_2,\cdots\mathbf{r}_N,t)\Psi(\mathbf{r}_1,\mathbf{r}_2,\cdots\mathbf{r}_N,t)$$

This last equation is in a very high dimension, so the solutions are not easy to visualize.

PROPERTIES

The Schrödinger equation has the following properties: some are useful, but there are shortcomings. Ultimately, these properties arise from the Hamiltonian used, and solutions to the equation.

Linearity

In the development above, the Schrödinger equation was made to be linear for generality, though this has other implications. If two wave functions ψ_1 and ψ_2 are solutions, then so is any linear combination of the two:

$$\psi = a\psi_1 + b\psi_2$$

where a and b are any complex numbers (the sum can be extended for any number of wavefunctions). This property allows superpositions of quantum states to be solutions of the Schrödinger equation. Even more generally, it holds that a general solution to the Schrödinger equation can be found by taking a weighted sum over all single state solutions achievable. For example, consider a wave function $\Psi(x, t)$ such that the wave function is a product of two functions: one time independent, and one time dependent. If states of definite energy found using the time independent Schrödinger equation are given by $\psi_E(x)$ with amplitude A_n and time dependent phase factor is given by

$$e^{-iE_n t/\hbar},$$

then a valid general solution is

$$\Psi(x,t) = \sum_n A_n \psi_{E_n}(x) e^{-iE_n t/\hbar}.$$

Additionally, the ability to scale solutions allows one to solve for a wave function without normalizing it first. If one has a set of normalized solutions ψ_n, then

$$\Psi = \sum_n A_n \psi_n$$

can be normalized by ensuring that

$$\sum_n |A_n|^2 = 1.$$

This is much more convenient than having to verify that

$$\int_{-\infty}^{\infty} |\Psi(x)|^2 \, dx = \int_{-\infty}^{\infty} \Psi(x)\Psi^*(x) \, dx = 1.$$

Real Energy Eigenstates

For the time-independent equation, an additional feature of linearity follows: if two wave functions ψ_1 and ψ_2 are solutions to the time-independent equation with the same energy E, then so is any linear combination:

$$\hat{H}(a\psi_1 + b\psi_2) = a\hat{H}\psi_1 + b\hat{H}\psi_2 = E(a\psi_1 + b\psi_2).$$

Two different solutions with the same energy are called degenerate.

In an arbitrary potential, if a wave function ψ solves the time-independent equation, so does its complex conjugate, denoted ψ*. By taking linear combinations, the real and imaginary parts of ψ are each solutions. If there is no degeneracy they can only differ by a factor.

In the time-dependent equation, complex conjugate waves move in opposite directions. If Ψ(x, t) is one solution, then so is Ψ(x, –t). The symmetry of complex conjugation is called time-reversal symmetry.

Space and Time Derivatives

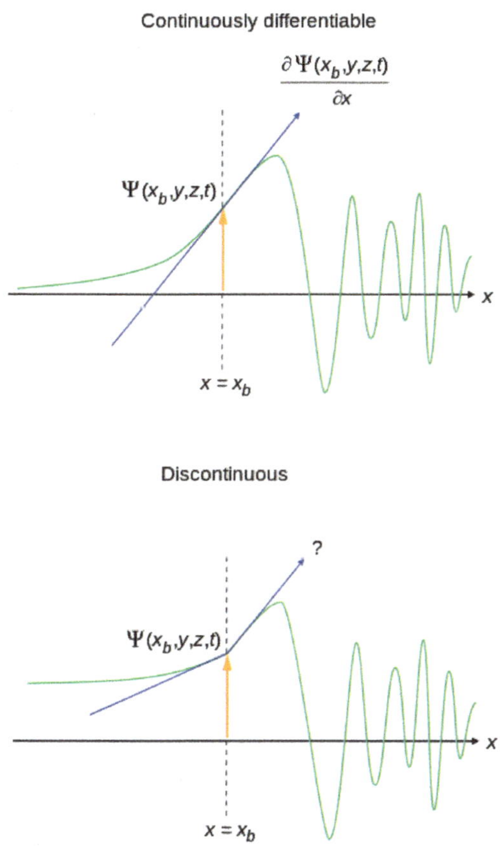

Continuity of the wavefunction and its first spatial derivative (in the x direction, yand z coordinates not shown), at some time t.

The Schrödinger equation is first order in time and second in space, which describes the time evolution of a quantum state (meaning it determines the future amplitude from the present).

Explicitly for one particle in 3-dimensional Cartesian coordinates – the equation is

$$i\hbar\frac{\partial\Psi}{\partial t}=-\frac{\hbar^2}{2m}\left(\frac{\partial^2\Psi}{\partial x^2}+\frac{\partial^2\Psi}{\partial y^2}+\frac{\partial^2\Psi}{\partial z^2}\right)+V(x,y,z,t)\Psi.$$

The first time partial derivative implies the initial value (at t = 0) of the wavefunction

$$\Psi(x,y,z,0)$$

is an arbitrary constant. Likewise – the second order derivatives with respect to space implies the wavefunction and its first order spatial derivatives

$$\Psi(x_b,y_b,z_b,t)$$

$$\frac{\partial}{\partial x}\Psi(x_b,y_b,z_b,t)\quad\frac{\partial}{\partial y}\Psi(x_b,y_b,z_b,t)\quad\frac{\partial}{\partial z}\Psi(x_b,y_b,z_b,t)$$

are all arbitrary constants at a given set of points, where x_b, y_b, z_b are a set of points describing boundary b (derivatives are evaluated at the boundaries). Typically there are one or two boundaries, such as the step potential and particle in a box respectively.

As the first order derivatives are arbitrary, the wavefunction can be a continuously differentiable function of space, since at any boundary the gradient of the wavefunction can be matched.

On the contrary, wave equations in physics are usually second order in time, notable are the family of classical wave equations and the quantum Klein–Gordon equation.

Local Conservation of Probability

The Schrödinger equation is consistent with probability conservation. Multiplying the Schrödinger equation on the right by the complex conjugate wavefunction, and multiplying the wavefunction to the left of the complex conjugate of the Schrödinger equation, and subtracting, gives the continuity equation for probability:

$$\frac{\partial}{\partial t}\rho(\mathbf{r},t)+\nabla\cdot\mathbf{j}=0,$$

where

$$\rho=|\Psi|^2=\Psi^*(\mathbf{r},t)\Psi(\mathbf{r},t)$$

is the probability density (probability per unit volume, * denotes complex conjugate), and

$$\mathbf{j}=\frac{1}{2m}(\Psi^*\hat{\mathbf{p}}\Psi-\Psi\hat{\mathbf{p}}\Psi^*)$$

is the probability current (flow per unit area).

Hence predictions from the Schrödinger equation do not violate probability conservation.

Positive Energy

If the potential is bounded from below, meaning there is a minimum value of potential energy, the eigenfunctions of the Schrödinger equation have energy which is also bounded from below. This can be seen most easily by using the variational principle, as follows. (See also below).

For any linear operator Â bounded from below, the eigenvector with the smallest eigenvalue is the vector ψ that minimizes the quantity

$$\langle\psi|\hat{A}|\psi\rangle$$

over all ψ which are normalized. In this way, the smallest eigenvalue is expressed through the variational principle. For the Schrödinger Hamiltonian Ĥ bounded from below, the smallest eigenvalue is called the ground state energy. That energy is the minimum value of

$$\langle\psi|\hat{H}|\psi\rangle = \int \psi^*(\mathbf{r})\left[-\frac{\hbar^2}{2m}\nabla^2\psi(\mathbf{r}) + V(\mathbf{r})\psi(\mathbf{r})\right]d^3\mathbf{r} = \int\left[\frac{\hbar^2}{2m}|\nabla\psi|^2 + V(\mathbf{r})|\psi|^2\right]d^3\mathbf{r} = \langle\hat{H}\rangle$$

(using integration by parts). Due to the complex modulus of ψ^2 (which is positive definite), the right hand side always greater than the lowest value of V(x). In particular, the ground state energy is positive when V(x) is everywhere positive.

For potentials which are bounded below and are not infinite over a region, there is a ground state which minimizes the integral above. This lowest energy wavefunction is real and positive definite – meaning the wavefunction can increase and decrease, but is positive for all positions. It physically cannot be negative: if it were, smoothing out the bends at the sign change (to minimize the wavefunction) rapidly reduces the gradient contribution to the integral and hence the kinetic energy, while the potential energy changes linearly and less quickly. The kinetic and potential energy are both changing at different rates, so the total energy is not constant, which can't happen (conservation). The solutions are consistent with Schrödinger equation if this wavefunction is positive definite.

The lack of sign changes also shows that the ground state is nondegenerate, since if there were two ground states with common energy E, not proportional to each other, there would be a linear combination of the two that would also be a ground state resulting in a zero solution.

Analytic Continuation to Diffusion

The above properties (positive definiteness of energy) allow the analytic continuation of the Schrödinger equation to be identified as a stochastic process. This can be interpreted as the Huygens–Fresnel principle applied to De Broglie waves; the spreading wavefronts are diffusive probability amplitudes. For a free particle (not subject to a potential) in a random walk, substituting τ = it into the time-dependent Schrödinger equation gives:

$$\frac{\partial}{\partial\tau}X(\mathbf{r},\tau) = \frac{\hbar}{2m}\nabla^2 X(\mathbf{r},\tau), \quad X(\mathbf{r},\tau) = \Psi(\mathbf{r},\tau/i)$$

which has the same form as the diffusion equation, with diffusion coefficient ħ/2m. In that case, the diffusivity yields the De Broglie relation in accordance with the Markov process.

Regularity

On the space L^2 of square-integrable densities, the Schrödinger semigroup $e^{it\hat{H}}$ is a unitary evolution, and therefore surjective. But the flows satisfy the Schrödinger equation $i\partial_t u = \widehat{H}u.$.

However, since $\widehat{H}$ for most physically reasonable Hamiltonians (e.g., the Laplace operator, possibly modified by a potential) is unbounded in L^2, this shows that the semigroup flows lack Sobolev regularity in general. Instead, solutions of the Schrödinger equation satisfy a Strichartz estimate.

RELATIVISTIC QUANTUM MECHANICS

Relativistic quantum mechanics is obtained where quantum mechanics and special relativity simultaneously apply. In general, one wishes to build relativistic wave equations from the relativistic energy–momentum relation

$$E^2 = (pc)^2 + (m_0c^2)^2 ,$$

instead of classical energy equations. The Klein–Gordon equation and the Dirac equation are two such equations. The Klein–Gordon equation,

$$\frac{1}{c^2}\frac{\partial^2}{\partial t^2}\psi - \nabla^2\psi + \frac{m^2c^2}{\hbar^2}\psi = 0.,$$

was the first such equation to be obtained, even before the nonrelativistic one, and applies to massive spinless particles. The Dirac equation arose from taking the "square root" of the Klein–Gordon equation by factorizing the entire relativistic wave operator into a product of two operators – one of these is the operator for the entire Dirac equation.

The general form of the Schrödinger equation remains true in relativity, but the Hamiltonian is less obvious. For example, the Dirac Hamiltonian for a particle of mass m and electric charge q in an electromagnetic field (described by the electromagnetic potentials φ and **A**) is:

$$\hat{H}_{\text{Dirac}} = \gamma^0 \left[c\boldsymbol{\gamma}\cdot(\hat{\mathbf{p}} - q\mathbf{A}) + mc^2 + \gamma^0 q\phi\right] ,$$

in which the $\gamma = (\gamma^1, \gamma^2, \gamma^3)$ and γ^0 are the Dirac gamma matrices related to the spin of the particle. The Dirac equation is true for all spin-$\frac{1}{2}$particles, and the solutions to the equation are 4-component spinor fields with two components corresponding to the particle and the other two for the antiparticle.

For the Klein–Gordon equation, the general form of the Schrödinger equation is inconvenient to use, and in practice the Hamiltonian is not expressed in an analogous way to the Dirac Hamiltonian. The equations for relativistic quantum fields can be obtained in other ways, such as starting from a Lagrangian density and using the Euler–Lagrange equations for fields, or use the representation theory of the Lorentz group in which certain representations can be used to fix the equation for a free particle of given spin (and mass).

In general, the Hamiltonian to be substituted in the general Schrödinger equation is not just a function of the position and momentum operators (and possibly time), but also of spin matrices. Also, the solutions to a relativistic wave equation, for a massive particle of spin s, are complex-valued 2(2s + 1)-component spinor fields.

QUANTUM FIELD THEORY

The general equation is also valid and used in quantum field theory, both in relativistic and nonrelativistic situations. However, the solution ψ is no longer interpreted as a wave, but should be interpreted as an operator acting on states existing in a Fock space.

FIRST ORDER FORM

The Schrödinger equation can also be derived from a first order form similar to the manner in which the Klein-Gordon equation can be derived from the Dirac equation. In 1D the first order equation is given by

$$-i\partial_z\psi = (i\eta\partial_t + \eta^\dagger m)\psi$$

This equation allows for the inclusion of spin in nonrelativistic quantum mechanics. Squaring the above equation yields the Schrödinger equation in 1D. The matrices η obey the following properties

$$\begin{aligned} \eta^2 &= 0 \\ (\eta^\dagger)^2 &= 0 \\ \{\eta, \eta^\dagger\} &= 2I \end{aligned}$$

The 3 dimensional version of the equation is given by

$$-i\gamma_i\partial_i\psi = (i\eta\partial_t + \eta^\dagger m)\psi$$

Here $\eta = (\gamma_0 + i\gamma_5)/\sqrt{2}$ is a 4×4 nilpotent matrix and γ_i are the Dirac gamma matrices ($i = 1,2,3$). The Schrödinger equation in 3D can be obtained by squaring the above equation. In the nonrelativistic limit $E - m \simeq E'$ and $E + m \simeq 2m$, the above equation can be derived from the Dirac equation.

REFEEENCES

1. Schrödinger, E. (1926). "An Undulatory Theory of the Mechanics of Atoms and Molecules" (PDF). Physical Review. **28** (6): 1049–1070. Bibcode:1926PhRv...28.1049S. doi:10.1103/PhysRev.28.1049. Archived from the original (PDF) on 17 December 2008.
2. Griffiths, David J. (2004), Introduction to Quantum Mechanics (2nd ed.), Prentice Hall, ISBN 0-13-111892-7
3. Laloe, Franck (2012), Do We Really Understand Quantum Mechanics, Cambridge University Press, ISBN 978-1-107-02501-1
4. Shankar, R. (1994). Principles of Quantum Mechanics (2nd ed.). Kluwer Academic/Plenum Publishers. ISBN 978-0-306-44790-7.
5. http://hyperphysics.phy-astr.gsu.edu/hbase/quantum/scheq.html
6. Sakurai, J. J. (1995). Modern Quantum Mechanics. Reading, Massachusetts: Addison-Wesley. p. 68.

7. Nouredine Zettili (17 February 2009). Quantum Mechanics: Concepts and Applications. John Wiley & Sons. ISBN 978-0-470-02678-6.
8. Ballentine, Leslie (1998), Quantum Mechanics: A Modern Development, World Scientific Publishing Co., ISBN 9810241054
9. David Deutsch, The Beginning of infinity, page 310
10. de Broglie, L. (1925). "Recherches sur la théorie des quanta" [On the Theory of Quanta] (PDF). Annales de Physique. **10** (3): 22–128. Translated version at the Wayback Machine (archived 9 May 2009).
11. Weissman, M.B.; V. V. Iliev; I. Gutman (2008). "A pioneer remembered: biographical notes about Arthur Constant Lunn". Communications in Mathematical and in Computer Chemistry. **59** (3): 687–708.
12. Kamen, Martin D. (1985). Radiant Science, Dark Politics. Berkeley and Los Angeles, CA: University of California Press. pp. 29–32. ISBN 0-520-04929-2.
13. Schrodinger, E. (1984). Collected papers. Friedrich Vieweg und Sohn. ISBN 3-7001-0573-8. See introduction to first 1926 paper.
14. Encyclopaedia of Physics (2nd Edition), R.G. Lerner, G.L. Trigg, VHC publishers, 1991, (Verlagsgesellschaft) 3-527-26954-1, (VHC Inc.) ISBN 0-89573-752-3
15. Sommerfeld, A. (1919). Atombau und Spektrallinien. Braunschweig: Friedrich Vieweg und Sohn. ISBN 3-87144-484-7.
16. For an English source, see Haar, T. "The Old Quantum Theory".
17. Rhodes, R. (1986). Making of the Atomic Bomb. Touchstone. ISBN 0-671-44133-7.
18. Erwin Schrödinger (1982). Collected Papers on Wave Mechanics: Third Edition. American Mathematical Soc. ISBN 978-0-8218-3524-1.
19. Schrödinger, E. (1926). "Quantisierung als Eigenwertproblem; von Erwin Schrödinger". Annalen der Physik. **384**: 361–377. Bibcode:1926AnP...384..361S. doi:10.1002/andp.19263840404.
20. Erwin Schrödinger, "The Present situation in Quantum Mechanics," p. 9 of 22. The English version was translated by John D. Trimmer. The translation first appeared first in **Proceedings of the American Philosophical Society**, 124, 323–38. It later appeared as Section I.11 of Part I of **Quantum Theory and Measurement** by J.A. Wheeler and W.H. Zurek, eds., Princeton University Press, New Jersey 1983.
21. Einstein, A.; et. al. "Letters on Wave Mechanics: Schrodinger–Planck–Einstein–Lorentz".
22. Moore, W.J. (1992). Schrödinger: Life and Thought. Cambridge University Press. ISBN 0-521-43767-9.
23. It is clear that even in his last year of life, as shown in a letter to Max Born, that Schrödinger never accepted the Copenhagen interpretation.[:220]
24. Takahisa Okino (2013). "Correlation between Diffusion Equation and Schrödinger Equation". Journal of Modern Physics (4): 612–615.

25. Molecular Quantum Mechanics Parts I and II: An Introduction to Quantum Chemistry (Volume 1), P.W. Atkins, Oxford University Press, 1977, ISBN 0-19-855129-0

26. The New Quantum Universe, T.Hey, P.Walters, Cambridge University Press, 2009, ISBN 978-0-521-56457-1

27. Quanta: A handbook of concepts, P.W. Atkins, Oxford University Press, 1974, ISBN 0-19-855493-1

28. Physics of Atoms and Molecules, B.H. Bransden, C.J.Joachain, Longman, 1983, ISBN 0-582-44401-2

29. Quantum Physics of Atoms, Molecules, Solids, Nuclei and Particles (2nd Edition), R. Resnick, R. Eisberg, John Wiley & Sons, 1985, ISBN 978-0-471-87373-0

30. Quantum Mechanics Demystified, D. McMahon, Mc Graw Hill (USA), 2006, ISBN 0-07-145546-9

31. Analytical Mechanics, L.N. Hand, J.D. Finch, Cambridge University Press, 2008, ISBN 978-0-521-57572-0

32. N. Zettili. Quantum Mechanics: Concepts and Applications (2nd ed.). p. 458. ISBN 978-0-470-02679-3.

33. Physical chemistry, P.W. Atkins, Oxford University Press, 1978, ISBN 0-19-855148-7

34. Solid State Physics (2nd Edition), J.R. Hook, H.E. Hall, Manchester Physics Series, John Wiley & Sons, 2010, ISBN 978-0-471-92804-1

35. Physics for Scientists and Engineers – with Modern Physics (6th Edition), P. A. Tipler, G. Mosca, Freeman, 2008, ISBN 0-7167-8964-7

36. David Griffiths (2008). Introduction to elementary particles. Wiley-VCH. pp. 162–. ISBN 978-3-527-40601-2. Retrieved 27 June 2011.

37. Quantum Mechanics, E. Abers, Pearson Ed., Addison Wesley, Prentice Hall Inc, 2004, ISBN 978-0-13-146100-0

38. http://www.stt.msu.edu/~mcubed/Relativistic.pdf

39. Takahisa Okino (2015). "Mathematical Physics in Diffusion Problems". Journal of Modern Physics (6): 2109–2144.

40. Ajaib, Muhammad Adeel (2015). "A Fundamental Form of the Schrödinger Equation". Found.Phys. 45 (2015) no.12, 1586-1598. doi:10.1007/s10701-015-9944-z.

41. Ajaib, Muhammad Adeel (2016). "Non-Relativistic Limit of the Dirac Equation". International Journal of Quantum Foundations.

42. Lévy-Leblond, J-.M. (1967). "Nonrelativistic particles and wave equations". Comm. Math. Pays. **6** (4): 286–311.

CHAPTER 8

Electronic Band Structure

Chapter Contents

- **Introduction**
- **Why Bands and Band Gaps Occur**
- **Theory of Band Structures in Crystals**
- **Band Diagrams**
- **Nearly Free Electron Model**
- **Muffin-Tin Approximation**
- **Density Functional Theory**
- **Empty Lattice Approximation**

A useful way to visualize the difference between conductors, insulators and semiconductors is to plot the available energies for electrons in the materials. Instead of having discrete energies as in the case of free atoms, the available energy states form bands. Crucial to the conduction process is whether or not there are electrons in the conduction band. In insulators the electrons in the valence band are separated by a large gap from the conduction band, in conductors like metals the valence band overlaps the conduction band, and in semiconductors there is a small enough gap between the valence and conduction bands that thermal or other excitations can bridge the gap. With such a small gap, the presence of a small percentage of a doping material can increase conductivity dramatically.

An important parameter in the band theory is the Fermi level, the top of the available electron energy levels at low temperatures. The position of the Fermi level with the relation to the conduction band is a crucial factor in determining electrical properties.

INTRODUCTION

In solid-state physics, the **electronic band structure** (or simply **band structure**) of a solid describes the range of energies that an electron within the solid may have (called energy bands, allowed bands, or simply bands) and ranges of energy that it may not have (called band gaps or forbidden bands).

Band theory derives these bands and band gaps by examining the allowed quantum mechanical wave functions for an electron in a large, periodic lattice of atoms or molecules. Band theory has been successfully used to explain many physical properties of solids, such as electrical resistivity and optical absorption, and forms the foundation of the understanding of all solid-state devices (transistors, solar cells, etc.).

WHY BANDS AND BAND GAPS OCCUR

The electrons of a single, isolated atom occupy atomic orbitals. Each orbital forms at a discrete energy level. When two atoms join together to form into a molecule, their atomic orbitals overlap. The Pauli exclusion principle dictates that no two electrons can have the same quantum numbers in a molecule. So if two identical atoms combine to form a diatomic molecule, each atomic orbital splits into two molecular orbitals of different energy, allowing the electrons in the former atomic orbitals to occupy the new orbital structure without any having the same energy

Similarly if a large number N of identical atoms come together to form a solid, such as a crystal lattice, the atoms› atomic orbitals overlap. Since the Pauli exclusion principle dictates that no two electrons in the solid have the same quantum numbers, each atomic orbital splits into N discrete molecular orbitals, each with a different energy. Since the number of atoms in a macroscopic piece of solid is a very large number ($N \sim 10^{22}$) the number of orbitals is very large and thus they are very closely spaced in energy. The energy of adjacent levels is so close together that they can be considered as a continuum, an energy band.

Band gaps are essentially leftover ranges of energy not covered by any band, a result of the finite widths of the energy bands. The bands have different widths, with the widths depending upon the degree of overlap in the atomic orbitals from which they arise. Two adjacent bands may simply not be wide enough to fully cover the range of energy. For example, the bands associated with core orbitals (such as 1s electrons) are extremely narrow due to the small overlap between adjacent atoms. As a result, there tend to be large band gaps between the core bands. Higher bands involve comparatively larger orbitals with more overlap, becoming progressively wider at higher energies so that there are no band gaps at higher energies.

Basic Concepts

Assumptions and Limits of Band Structure Theory

To start out, it is important to note what has been assumed in order to gain the simplicity of the band theory:

- Infinite-size system: For the bands to be continuous, we must consider a large piece of material. The concept of band structure can be extended to systems which are only "large" along reduced dimensions, such as two-dimensional electron systems.
- Homogeneous system: The notion of a band structure as an intrinsic property of a material assumes that the material is homogeneous in some way. Practically, this means that band structure describes the bulk inside a uniform piece of material.
- Non-interactivity: The band structure describes "single electron states". The existence of these states assumes that the electrons travel in a static potential without dynamically interacting with lattice vibrations, other electrons, photons, etc.

The above assumptions are broken in a number of important practical situations, and the use of band structure requires one to keep a close check on the limitations of band theory:

- Inhomogeneities and interfaces: Near surfaces, junctions, and other inhomogeneities, the bulk band structure is disrupted. Not only are there local small-scale disruptions (e.g., surface states or dopant states inside the band gap), but also local charge imbalances. These charge imbalances have electrostatic effects that extend deeply into semiconductors, insulators, and the vacuum (see doping, band bending).
- Along the same lines, most electronic effects (capacitance, electrical conductance, electric-field screening) involve the physics of electrons passing through surfaces and/or near interfaces. The full description of these effects, in a band structure picture, requires at least a rudimentary model of electron-electron interactions (see space charge, band bending).
- Small systems: For systems which are small along every dimension (e.g., a small molecule or a quantum dot), there is no continuous band structure. The crossover between small and large dimensions is the realm of mesoscopic physics.
- Strongly correlated materials (for example, Mott insulators) simply cannot be understood in terms of single-electron states. The electronic band structures of these materials are poorly defined (or at least, not uniquely defined) and may not provide useful information about their physical state.

Crystalline Symmetry and Wavevectors

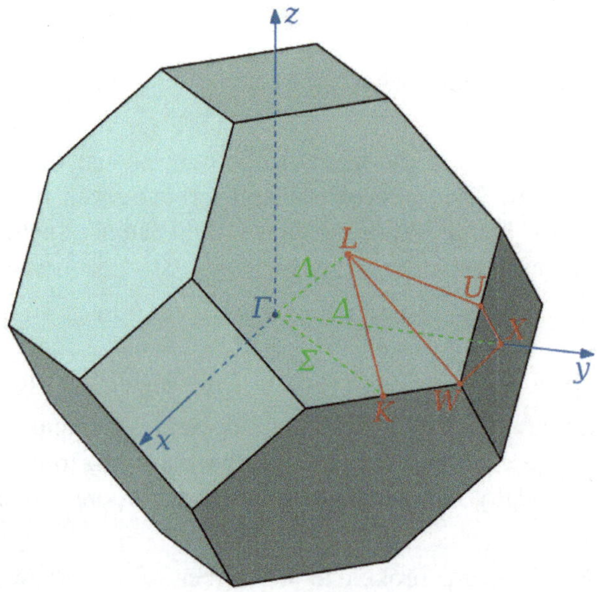

Fig 1: Brillouin zone of a face-centered cubic lattice showing labels for special symmetry points.

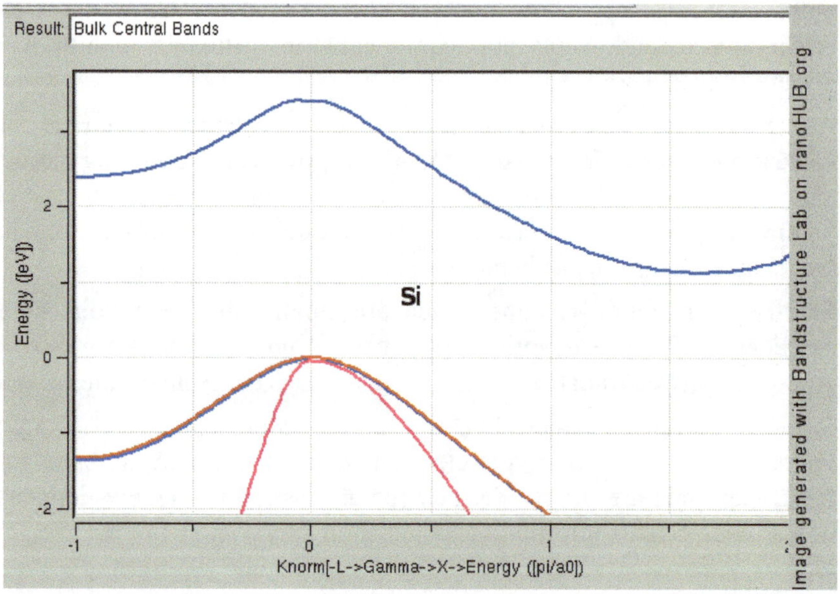

Fig 2. Band structure plot for Si, Ge, GaAs and InAs generated with tight binding model. Note that Si and Ge are indirect band gap materials, while GaAs and InAs are direct.

Band structure calculations take advantage of the periodic nature of a crystal lattice, exploiting its symmetry. The single-electron Schrödinger equation is solved for an electron in a lattice-periodic potential, giving Bloch waves as solutions:

$$\psi_{n\mathbf{k}}(\mathbf{r}) = e^{i\mathbf{k}\cdot\mathbf{r}} u_{n\mathbf{k}}(\mathbf{r}),$$

where **k** is called the wavevector. For each value of **k**, there are multiple solutions to the Schrödinger equation labelled by n, the band index, which simply numbers the energy bands. Each of these energy levels evolves smoothly with changes in **k**, forming a smooth band of states. For each band we can define a function $E_n(\mathbf{k})$, which is the dispersion relation for electrons in that band.

The wavevector takes on any value inside the Brillouin zone, which is a polyhedron in wavevector space that is related to the crystal›s lattice. Wavevectors outside the Brillouin zone simply correspond to states that are physically identical to those states within the Brillouin zone. Special high symmetry points/lines in the Brillouin zone are assigned labels like Γ, Δ, Λ, Σ (see Fig 1).

It is difficult to visualize the shape of a band as a function of wavevector, as it would require a plot in four-dimensional space, E vs. k_x, k_y, k_z. In scientific literature it is common to see **band structure plots** which show the values of $E_n(\mathbf{k})$ for values of **k** along straight lines connecting symmetry points, often labelled Δ, Λ, Σ, or , , and , respectively. Another method for visualizing band structure is to plot a constant-energy isosurface in wavevector space, showing all of the states with energy equal to a particular value. The isosurface of states with energy equal to the Fermi level is known as the Fermi surface.

Energy band gaps can be classified using the wavevectors of the states surrounding the band gap:

- Direct band gap: the lowest-energy state above the band gap has the same **k** as the highest-energy state beneath the band gap.
- Indirect band gap: the closest states above and beneath the band gap do not have the same **k** value.

Asymmetry: Band Structures in non-crystalline Solids

Although electronic band structures are usually associated with crystalline materials, quasi-crystalline and amorphous solids may also exhibit band structures. These are somewhat more difficult to study theoretically since they lack the simple symmetry of a crystal, and it is not usually possible to determine a precise dispersion relation. As a result, virtually all of the existing theoretical work on the electronic band structure of solids has focused on crystalline materials.

Density of States

The density of states function g(E) is defined as the number of electronic states per unit volume, per unit energy, for electron energies near E.

The density of states function is important for calculations of effects based on band theory. It appears in calculations for optical absorption where it provides both the number of excitable electrons and the number of final states for an electron. It appears in calculations of electrical conductivity where it provides the number of mobile states, and in computing electron scattering rates where it provides the number of final states after scattering.

For energies inside a band gap, g(E) = 0.

Filling of Bands

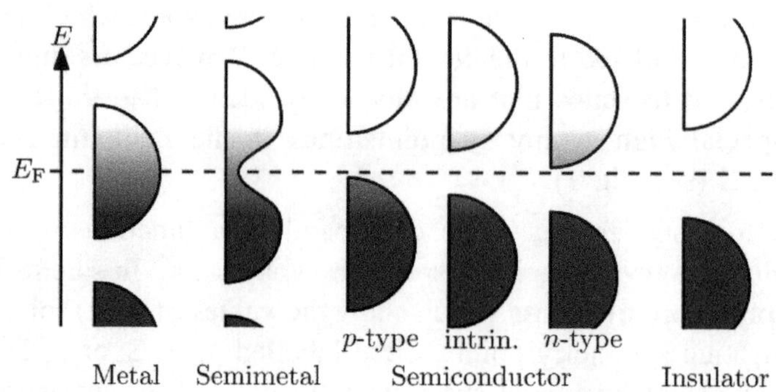

Filling of the electronic states in various types of materials at equilibrium. Here, height is energy while width is the density of available states for a certain energy in the material listed. The shade follows the Fermi–Dirac distribution (**black** = all states filled, **white** = no state filled). In metals and semimetals the Fermi level E_F lies inside at least one band. In insulators and semiconductors the Fermi level is inside a band gap; however, in semiconductors the bands are near enough to the Fermi level to be thermally populated with electrons or holes.

At thermodynamic equilibrium, the likelihood of a state of energy E being filled with an electron is given by the Fermi–Dirac distribution, a thermodynamic distribution that takes into account the Pauli exclusion principle:

$$f(E) = \frac{1}{1 + e^{(E-\mu)/k_B T}}$$

where:

- $k_B T$ is the product of Boltzmann›s constant and temperature, and
- μ is the total chemical potential of electrons, or Fermi level (in semiconductor physics, this quantity is more often denoted E_F). The Fermi level of a solid is directly related to the voltage on that solid, as measured with a voltmeter. Conventionally, in band structure plots the Fermi level is taken to be the zero of energy (an arbitrary choice).

The density of electrons in the material is simply the integral of the Fermi–Dirac distribution times the density of states:

$$N/V = \int_{-\infty}^{\infty} g(E)f(E)\,dE$$

Although there are an infinite number of bands and thus an infinite number of states, there are only a finite number of electrons to place in these bands. The preferred value for the number of electrons is a consequence of electrostatics: even though the surface of a material can be charged, the internal bulk of a material prefers to be charge neutral. The condition of charge neutrality means that N/V must match the density of protons in the material. For this to occur, the material electrostatically adjusts itself, shifting its band structure up or down in energy (thereby shifting g(E)), until it is at the correct equilibrium with respect to the Fermi level.

Names of bands near the Fermi level (Conduction band, Valence Band)

A solid has an infinite number of allowed bands, just as an atom has infinitely many energy levels. However, most of the bands simply have too high energy, and are usually disregarded under ordinary circumstances. Conversely, there are very low energy bands associated with the core orbitals (such as 1s electrons). These low-energy core bands are also usually disregarded since they remain filled with electrons at all times, and are therefore inert. Likewise, materials have several band gaps throughout their band structure.

The most important bands and band gaps—those relevant for electronics and optoelectronics—are those with energies near the Fermi level. The bands and band gaps near the Fermi level are given special names, depending on the material:

- In a semiconductor or band insulator, the Fermi level is surrounded by a band gap, referred to as the band gap (to distinguish it from the other band gaps in the band structure). The closest band above the band gap is called the conduction band, and the closest band beneath the band gap is called the valence band. The name "valence band" was coined by analogy to chemistry, since in many semiconductors the valence band is built out of the valence orbitals.
- In a metal or semimetal, the Fermi level is inside of one or more allowed bands. In semimetals the bands are usually referred to as "conduction band" or "valence band" depending on whether the charge transport is more electron-like or hole-like, by analogy to semiconductors. In many metals, however, the bands are neither electron-like nor hole-like, and often just called "valence band" as they are made of valence orbitals. The band gaps in a metal's band structure are not important for low energy physics, since they are too far from the Fermi level.

THEORY OF BAND STRUCTURES IN CRYSTALS

The ansatz is the special case of electron waves in a periodic crystal lattice using Bloch waves as treated generally in the dynamical theory of diffraction. Every crystal is a periodic structure which can be characterized by a Bravais lattice, and for each Bravais lattice we can determine the reciprocal lattice, which encapsulates the periodicity in a set of three reciprocal lattice vectors ($\mathbf{b}_1$,$\mathbf{b}_2$,$\mathbf{b}_3$). Now, any periodic potential V(**r**) which shares the same periodicity as the direct lattice can be expanded out as a Fourier series whose only non-vanishing components are those associated with the reciprocal lattice vectors. So the expansion can be written as:

$$V(\mathbf{r}) = \sum_{\mathbf{K}} V_{\mathbf{K}} e^{i\mathbf{K}\cdot\mathbf{r}}$$

where $\mathbf{K} = m_1\mathbf{b}_1 + m_2\mathbf{b}_2 + m_3\mathbf{b}_3$ for any set of integers (m_1,m_2,m_3).

From this theory, an attempt can be made to predict the band structure of a particular material, however most ab initio methods for electronic structure calculations fail to predict the observed band gap.

Nearly free Electron Approximation

In the nearly free electron approximation, interactions between electrons are completely ignored. This approximation allows use of Bloch's Theorem which states that electrons in a periodic potential have wavefunctions and energies which are periodic in wavevector up to a constant phase shift between neighboring reciprocal lattice vectors. The consequences of periodicity are described mathematically by the Bloch wavefunction:

$$\Psi_{n,\mathbf{k}}(\mathbf{r}) = e^{i\mathbf{k}\cdot\mathbf{r}} u_n(\mathbf{r})$$

where the function $u_n(\mathbf{r})$ is periodic over the crystal lattice, that is,

$$u_n(\mathbf{r}) = u_n(\mathbf{r} - \mathbf{R}).$$

Here index n refers to the n-th energy band, wavevector **k** is related to the direction of motion of the electron, **r** is the position in the crystal, and **R** is the location of an atomic site.

The NFE model works particularly well in materials like metals where distances between neighbouring atoms are small. In such materials the overlap of atomic orbitals and potentials on neighbouring atoms is relatively large. In that case the wave function of the electron can be approximated by a (modified) plane wave. The band structure of a metal like Aluminium even gets close to the empty lattice approximation.

Tight Binding Model

The opposite extreme to the nearly free electron approximation assumes the electrons in the crystal behave much like an assembly of constituent atoms. This tight binding model assumes the solution to the time-independent single electron Schrödinger equation Ψ is well approximated by a linear combination of atomic orbitals $\psi_n(\mathbf{r})$.

$$\Psi(\mathbf{r}) = \sum_{n,\mathbf{R}} b_{n,\mathbf{R}} \psi_n(\mathbf{r} - \mathbf{R}),$$

where the coefficients $b_{n,\mathbf{R}}$ are selected to give the best approximate solution of this form. Index n refers to an atomic energy level and **R** refers to an atomic site. A more accurate approach using this idea employs Wannier functions, defined by:

$$a_n(\mathbf{r}-\mathbf{R}) = \frac{V_C}{(2\pi)^3} \int_{\mathrm{BZ}} d\mathbf{k} e^{-i\mathbf{k}\cdot(\mathbf{R}-\mathbf{r})} u_{n\mathbf{k}};$$

in which $u_{n\mathbf{k}}$ is the periodic part of the Bloch wave and the integral is over the Brillouin zone. Here index n refers to the n-th energy band in the crystal. The Wannier functions are localized near atomic sites, like atomic orbitals, but being defined in terms of Bloch functions they are accurately related to solutions based upon the crystal potential. Wannier functions on different atomic sites **R** are orthogonal. The Wannier functions can be used to form the Schrödinger solution for the n-th energy band as:

$$\Psi_{n,\mathbf{k}}(\mathbf{r}) = \sum_{\mathbf{R}} e^{-i\mathbf{k}\cdot(\mathbf{R}-\mathbf{r})} a_n(\mathbf{r}-\mathbf{R}).$$

The TB model works well in materials with limited overlap between atomic orbitals and potentials on neighbouring atoms. Band structures of materials like Si, GaAs, SiO_2 and diamond for instance are well described by TB-Hamiltonians on the basis of atomic sp^3 orbitals. In transition metals a mixed TB-NFE model is used to describe the broad NFE conduction band and the narrow embedded TB d-bands. The radial functions of the atomic orbital part of the Wannier functions are most easily calculated by the use of pseudopotential methods. NFE, TB or combined NFE-TB band structure calculations, sometimes extended with wave function approximations based on pseudopotential methods, are often used as an economic starting point for further calculations.

KKR Model

The simplest form of this approximation centers non-overlapping spheres (referred to as muffin tins) on the atomic positions. Within these regions, the potential experienced by an electron is approximated to be spherically symmetric about the given nucleus. In the remaining interstitial region, the screened potential is approximated as a constant. Continuity of the potential between the atom-centered spheres and interstitial region is enforced.

A variational implementation was suggested by Korringa and by Kohn and Rostocker, and is often referred to as the KKR model.

Density-Functional Theory

In recent physics literature, a large majority of the electronic structures and band plots are calculated using density-functional theory (DFT), which is not a model but rather a theory, i.e., a microscopic first-principles theory of condensed matter physics that tries to cope with the electron-electron many-body problem via the introduction of an exchange-correlation term in the functional of the electronic density. DFT-calculated bands are in many cases found to be in agreement with experimentally measured bands, for example by angle-resolved photoemission spectroscopy (ARPES). In particular, the band shape is typically well reproduced by DFT. But there are also systematic errors in DFT bands when compared to experiment results. In particular, DFT seems to systematically underestimate by about 30-40% the band gap in insulators and semiconductors.

It is commonly believed that DFT is a theory to predict ground state properties of a system only (e.g. the total energy, the atomic structure, etc.), and that excited state properties cannot be determined by DFT. This is a misconception. In principle, DFT can determine any property (ground state or excited state) of a system given a functional that maps the ground state density to that property. This is the essence of the Hohenburg–Kohn theorem. In practice, however, no known functional exists that maps the ground state density to excitation energies of electrons within a material. Thus, what in the literature is quoted as a DFT band plot is a representation of the DFT Kohn–Sham energies, i.e., the energies of a fictive non-interacting system, the Kohn–Sham system, which has no physical interpretation at all. The Kohn–Sham electronic structure must not be confused with the real, quasiparticle electronic structure of a system, and there is no Koopman›s theorem holding for Kohn–Sham energies, as there is for Hartree–Fock energies, which can be truly considered as an approximation for quasiparticle energies. Hence, in principle, Kohn–Sham based DFT is not a band theory, i.e., not a theory suitable for calculating bands and band-plots. In principle time-dependent DFT can be used to calculate the true band structure although in practice this is often difficult. A popular approach is the use of hybrid functionals, which incorporate a portion of Hartree–Fock exact exchange; this produces a substantial improvement in predicted bandgaps of semiconductors, but is less reliable for metals and wide-bandgap materials.

Green's Function Methods and the *ab initio* GW approximation

To calculate the bands including electron-electron interaction many-body effects, one can resort to so-called Green›s function methods. Indeed, knowledge of the Green's function of a system provides both ground (the total energy) and also excited state observables of the system. The poles of the Green's function are the quasiparticle energies, the bands of a solid. The Green's function can be calculated by solving the Dyson equation once the self-energy of the system is known. For real systems like solids, the self-energy is a very complex quantity and usually approximations are needed to solve the problem. One such approximation is the GW approximation, so called from the mathematical form the self-energy takes as the product $\Sigma = GW$ of the Green's function G and the dynamically screened interaction W. This approach is more pertinent when addressing the calculation of band plots (and also quantities beyond, such as the spectral function) and can also be formulated in a completely ab initio way. The GW approximation seems to provide band gaps of insulators and semiconductors in agreement with experiment, and hence to correct the systematic DFT underestimation.

Mott insulators

Although the nearly free electron approximation is able to describe many properties of electron band structures, one consequence of this theory is that it predicts the same number of electrons in each unit cell. If the number of electrons is odd, we would then expect that there is an unpaired electron in each unit cell, and thus that the valence band is not fully occupied, making the material a conductor. However, materials such as CoO that have an odd number of electrons per unit cell are insulators, in direct conflict with this result. This kind of material is known as a Mott insulator, and requires inclusion of detailed electron-electron interactions (treated only as an averaged effect on the crystal potential in band theory) to explain the discrepancy. The Hubbard model is an approximate theory that can include these interactions. It can be treated non-perturbatively within the so-called dynamical mean field theory, which attempts to bridge the gap between the nearly free electron approximation and the atomic limit. Formally, however, the states are not non-interacting in this case and the concept of a band structure is not adequate to describe these cases.

Others

Calculating band structures is an important topic in theoretical solid state physics. In addition to the models mentioned above, other models include the following:

- Empty lattice approximation: the "band structure" of a region of free space that has been divided into a lattice.
- k·p perturbation theory is a technique that allows a band structure to be approximately described in terms of just a few parameters. The technique is commonly used for semiconductors, and the parameters in the model are often determined by experiment.
- The Kronig-Penney Model, a one-dimensional rectangular well model useful for illustration of band formation. While simple, it predicts many important phenomena, but is not quantitative.
- Hubbard model

The band structure has been generalised to wavevectors that are complex numbers, resulting in what is called a complex band structure, which is of interest at surfaces and interfaces.

Each model describes some types of solids very well, and others poorly. The nearly free electron model works well for metals, but poorly for non-metals. The tight binding model is extremely accurate for ionic insulators, such as metal halide salts (e.g. NaCl).

BAND DIAGRAMS

To understand how band structure changes relative to the Fermi level in real space, a band structure plot is often first simplified in the form of a band diagram. In a band diagram the vertical axis is energy while the horizontal axis represents real space. Horizontal lines represent energy levels, while blocks represent energy bands. When the horizontal lines in these diagram are slanted then the energy of the level or band changes with distance. Diagrammatically, this depicts the presence of an electric field within the crystal system. Band diagrams are useful in relating the general band structure properties of different materials to one another when placed in contact with each other.

NEARLY FREE ELECTRON MODEL

In solid-state physics, the **nearly free electron model** (or **NFE model**) is a quantum mechanical model of physical properties of electrons that can move almost freely through the crystal lattice of a solid. The model is closely related to the more conceptual Empty Lattice Approximation. The model enables understanding and calculating the electronic band structure of especially metals.

Introduction - a Heuristic Argument

Free electrons are traveling plane waves. Generally the time independent part of their wave function is expressed as

$$\psi_{\mathbf{k}}(\mathbf{r}) = \frac{1}{\sqrt{\Omega_r}} e^{i\mathbf{k}\cdot\mathbf{r}}$$

These plane wave solutions have an energy of

$$E_k = \frac{\hbar^2 k^2}{2m}$$

The expression of the plane wave as a complex exponential function can also be written as the sum of two periodic functions which are mutually shifted a quarter of a period.

$$\psi_{\mathbf{k}}(\mathbf{r}) = \frac{1}{\sqrt{\Omega_r}} [\cos(\mathbf{k}\cdot\mathbf{r}) + i\sin(\mathbf{k}\cdot\mathbf{r})]$$

In this light the wave function of a free electron can be viewed as the sum of two plane waves. Sine and cosine functions can also be expressed as sums or differences of plane waves moving in opposite directions

$$\cos(\mathbf{k}\cdot\mathbf{r}) = \frac{1}{2}[e^{i\mathbf{k}\cdot\mathbf{r}} + e^{-i\mathbf{k}\cdot\mathbf{r}}]$$

Assume that there is only one kind of atom present in the lattice and that the atoms are located at the lattice points. The potential of the atoms is attractive (negative) and concentrated near the lattice points. In the remainder of the cell the potential is close to zero.

The Hamiltonian is expressed as

$$H = T + V = -\frac{\hbar^2}{2m}\nabla^2 + V(\mathbf{r})$$

in which {\displaystyle T} is the kinetic and V is the potential energy. From this expression the energy expectation value, or the statistical average, of the energy of the electron can be calculated with

$$E = \langle H \rangle = \int_{\Omega_r} \psi_{\mathbf{k}}^*(\mathbf{r})[T+V]\psi_{\mathbf{k}}(\mathbf{r})d\mathbf{r}$$

where we integrate in r over a single lattice cell. If we assume that the electron is given by a plane wave of wave number k (despite the nonconstant potential V, the energy of the electron is:

$$E_k = \frac{1}{\Omega_r}\int_{\Omega_r} e^{-i\mathbf{k}\cdot\mathbf{r}}\left[\frac{\hbar^2 k^2}{2m} + V(\mathbf{r})\right] e^{i\mathbf{k}\cdot\mathbf{r}} d\mathbf{r} = \frac{\hbar^2 k^2}{2m} + \langle V \rangle$$

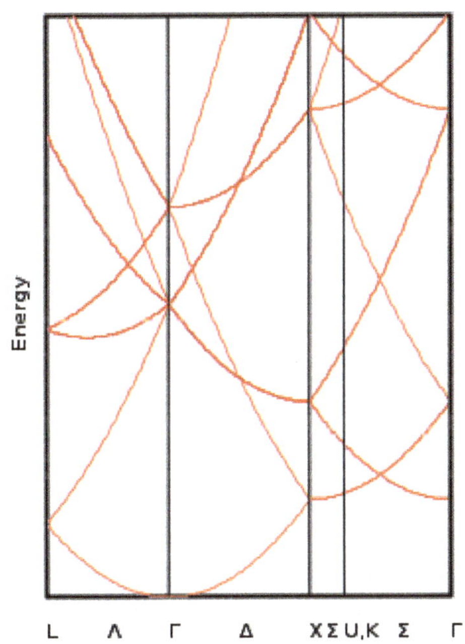

Some free electron bands in a FCC crystal structure according to the **Empty Lattice Approximation**

This means that at each k the energy is lowered below the free space value by the average Vof the attractive potential of the atom. If the potential is very small we get the Empty Lattice Approximation. This isn›t a very sensational result and it doesn›t say anything about what happens when we get close to the Brillouin zone boundary. We will look at those regions in $\mathbf{k}$-space now.

Let's assume that we look at the problem from the origin, at position $\mathbf{r} = 0$. If $\mathbf{k} = 0$ only the cosine part is present and the sine part is moved to ∞. If we let the length of the wave vector k grow, then the central maximum of the cosine part stays at $\mathbf{r} = 0$. The first maximum and minimum of the sine part are at $\mathbf{r} = \pm\pi/(2\mathbf{k})$. They come nearer as k grows. Let's assume that k is close to the Brillouin zone boundary for the analysis in the next part of this introduction.

The atomic positions coincide with the maximum of the $\cos(\mathbf{k}\cdot\mathbf{r})$-component of the wave function. The interaction of the $\cos(\mathbf{k}\cdot\mathbf{r})$-component of the wave function with the potential will be different from the interaction of the $\rho_{\mathbf{k}}$-component of the wave function with the potential because their phases are shifted. The charge density $|\psi_{\mathbf{k}}(\mathbf{r})|^2$ is proportional to the absolute value squared, $\rho_{\mathbf{k}} = \rho^c_{\mathbf{k}}(\mathbf{r}) + \rho^s_{\mathbf{k}}(\mathbf{r})$, of the wave function. It is useful to split it into two parts, $\rho_{\mathbf{k}} = \rho^c_{\mathbf{k}}(\mathbf{r}) + \rho^s_{\mathbf{k}}(\mathbf{r})$, coming from the $\cos(\mathbf{k}\cdot\mathbf{r})$, and $\sin(\mathbf{k}\cdot\mathbf{r})$ -components. For the former component it is

$$\rho^c_{\mathbf{k}}(\mathbf{r}) = \frac{1}{2\Omega}\left[1 + \cos(2\mathbf{k}\cdot\mathbf{r})\right]$$

and for the $\sin(\mathbf{k}\cdot\mathbf{r})$-component it is

$$\rho^s_{\mathbf{k}}(\mathbf{r}) = \frac{1}{2\Omega}\left[1 - \cos(2\mathbf{k}\cdot\mathbf{r})\right]$$

For values of k close to the Brillouin zone boundary, the length of the two waves and the period of the two different charge density distributions almost coincide with the periodic potential of the lattice. As a result the charge densities of the two components have a different energy because the maximum of the charge density of the $\cos(\mathbf{k}\cdot\mathbf{r})$ -component coincides with the attractive potential of the atoms while the maximum of the charge density of the $\sin(\mathbf{k}\cdot\mathbf{r})$-component lies in the regions with a higher electrostatic potential between the atoms.

As a result the aggregate will be split in high and low energy components when the kinetic energy increases and the wave vector approaches the length of the reciprocal lattice vectors. The potentials of the atomic cores can be decomposed into Fourier components to meet the requirements of a description in terms of reciprocal space parameters.

Mathematical Formulation

The nearly free electron model is a modification of the free-electron gas model which includes a weak periodic perturbation meant to model the interaction between the conduction electrons and the ions in a crystalline solid. This model, like the free-electron model, does not take into account electron-electron interactions; that is, the independent-electron approximation is still in effect.

As shown by Bloch›s theorem, introducing a periodic potential into the Schrödinger equation results in a wave function of the form

$$\psi_{\mathbf{k}}(\mathbf{r}) = u_{\mathbf{k}}(\mathbf{r})e^{i\mathbf{k}\cdot\mathbf{r}}$$

where the function *u* has the same periodicity as the lattice:

$$u_{\mathbf{k}}(\mathbf{r}) = u_{\mathbf{k}}(\mathbf{r} + \mathbf{T})$$

(where *T* is a lattice translation vector.)

Because it is a *nearly* free electron approximation we can assume that

$$u_{\mathbf{k}}(\mathbf{r}) \approx \frac{1}{\sqrt{\Omega_r}}$$

A solution of this form can be plugged into the Schrödinger equation, resulting in the **central equation**:

$$(\lambda_{\mathbf{k}} - \epsilon)C_{\mathbf{k}} + \sum_{\mathbf{G}} U_{\mathbf{G}} C_{\mathbf{k}-\mathbf{G}} = 0$$

where the kinetic energy $\lambda_{\mathbf{k}}$ is given by

$$\lambda_{\mathbf{k}}\psi_{\mathbf{k}}(\mathbf{r}) = -\frac{\hbar^2}{2m}\nabla^2\psi_{\mathbf{k}}(\mathbf{r}) = -\frac{\hbar^2}{2m}\nabla^2(u_{\mathbf{k}}(\mathbf{r})e^{i\mathbf{k}\cdot\mathbf{r}})$$

which, after dividing by $\psi_{\mathbf{k}}(\mathbf{r})$, reduces to

$$\lambda_{\mathbf{k}} = \frac{\hbar^2 k^2}{2m}$$

if we assume that $u_{\mathbf{k}}(\mathbf{r})$ is almost constant and $\nabla^2 u_{\mathbf{k}}(\mathbf{r}) \ll k^2$

The reciprocal parameters C_k and U_G are the Fourier coefficients of the wave function $\psi(r)$ and the screened potential energy U(r), respectively:

$$U(\mathbf{r}) = \sum_{\mathbf{G}} U_{\mathbf{G}} e^{i\mathbf{G}\cdot\mathbf{r}}$$

$$\psi(\mathbf{r}) = \sum_{\mathbf{k}} C_{\mathbf{k}} e^{i\mathbf{k}\cdot\mathbf{r}}$$

The vectors G are the reciprocal lattice vectors, and the discrete values of k are determined by the boundary conditions of the lattice under consideration.

In any perturbation analysis, one must consider the base case to which the perturbation is applied. Here, the base case is with U(x) = 0, and therefore all the Fourier coefficients of the potential are also zero. In this case the central equation reduces to the form

$$(\lambda_{\mathbf{k}} - \epsilon)C_{\mathbf{k}} = 0$$

This identity means that for each k, one of the two following cases must hold:

1. $C_{\mathbf{k}} = 0$,
2. $\lambda_{\mathbf{k}} = \epsilon$

If the values of λ_k are non-degenerate, then the second case occurs for only one value of k, while for the rest, the Fourier expansion coefficient C_k must be zero. In this non-degenerate case, the standard free electron gas result is retrieved:

$$\psi_k \propto e^{i\mathbf{k}\cdot\mathbf{r}}$$

In the degenerate case, however, there will be a set of lattice vectors k_1, ..., k_m with $\lambda_1 = ... = \lambda_m$. When the energy ϵ is equal to this value of λ, there will be m independent plane wave solutions of which any linear combination is also a solution:

$$\psi \propto \sum_{i=1}^{m} A_i e^{i\mathbf{k}_i\cdot\mathbf{r}}$$

Non-degenerate and degenerate perturbation theory can be applied in these two cases to solve for the Fourier coefficients C_k of the wavefunction (correct to first order in U) and the energy eigenvalue (correct to second order in U). An important result of this derivation is that there is no first-order shift in the energy ε in the case of no degeneracy, while there is in the case of near-degeneracy, implying that the latter case is more important in this analysis. Particularly, at the Brillouin zone boundary (or, equivalently, at any point on a Bragg plane), one finds a twofold energy degeneracy that results in a shift in energy given by:

$$\epsilon = \lambda_{\mathbf{k}} \pm |U_{\mathbf{k}}|$$

This **energy gap** between Brillouin zones is known as the band gap, with a magnitude of $2|U_K|$.

Results

Introducing this weak perturbation has significant effects on the solution to the Schrödinger equation, most significantly resulting in a band gap between wave vectors in different Brillouin zones.

Justifications

In this model, the assumption is made that the interaction between the conduction electrons and the ion cores can be modeled through the use of a "weak" perturbing potential. This may seem like a severe approximation, for the Coulomb attraction between these two particles of opposite charge can be quite significant at short distances. It can be partially justified, however, by noting two important properties of the quantum mechanical system:

- The force between the ions and the electrons is greatest at very small distances. However, the conduction electrons are not "allowed" to get this close to the ion cores due to the Pauli exclusion principle: the orbitals closest to the ion core are already occupied by the core electrons. Therefore, the conduction electrons never get close enough to the ion cores to feel their full force.

- Furthermore, the core electrons shield the ion charge magnitude “seen” by the conduction electrons. The result is an effective nuclear charge experienced by the conduction electrons which is significantly reduced from the actual nuclear charge.

Tight Binding

In solid-state physics, the **tight-binding model** (or **TB model**) is an approach to the calculation of electronic band structure using an approximate set of wave functions based upon superposition of wave functions for isolated atoms located at each atomic site. The method is closely related to the LCAO method (linear combination of atomic orbitals method) used in chemistry. Tight-binding models are applied to a wide variety of solids. The model gives good qualitative results in many cases and can be combined with other models that give better results where the tight-binding model fails. Though the tight-binding model is a one-electron model, the model also provides a basis for more advanced calculations like the calculation of surface states and application to various kinds of many-body problem and quasiparticle calculations.

The name “tight binding” of this electronic band structure model suggests that this quantum mechanical model describes the properties of tightly bound electrons in solids. The electrons in this model should be tightly bound to the atom to which they belong and they should have limited interaction with states and potentials on surrounding atoms of the solid. As a result the wave function of the electron will be rather similar to the atomic orbital of the free atom to which it belongs. The energy of the electron will also be rather close to the ionization energy of the electron in the free atom or ion because the interaction with potentials and states on neighboring atoms is limited.

Though the mathematical formulation of the one-particle tight-binding Hamiltonian may look complicated at first glance, the model is not complicated at all and can be understood intuitively quite easily. There are only three kinds of matrix elements that play a significant role in the theory. Two of those three kinds of elements should be close to zero and can often be neglected. The most important elements in the model are the interatomic matrix elements, which would simply be called the bond energies by a chemist.

In general there are a number of atomic energy levels and atomic orbitals involved in the model. This can lead to complicated band structures because the orbitals belong to different point-group representations. The reciprocal lattice and the Brillouin zone often belong to a different space group than the crystal of the solid. High-symmetry points in the Brillouin zone belong to different point-group representations. When simple systems like the lattices of elements or simple compounds are studied it is often not very difficult to calculate eigenstates in high-symmetry points analytically. So the tight-binding model can provide nice examples for those who want to learn more about group theory.

The tight-binding model has a long history and has been applied in many ways and with many different purposes and different outcomes. The model doesn›t stand on its own. Parts of the model can be filled in or extended by other kinds of calculations and models like the nearly-free electron model. The model itself, or parts of it, can serve as the basis for other calculations. In the study of conductive polymers, organic semiconductors and molecular electronics, for example, tight-binding-like models are applied in which the role of the atoms in the original concept is replaced by the molecular orbitals of conjugated systems and where the interatomic matrix elements are replaced by inter- or intramolecular hopping and tunneling parameters. These conductors nearly all have very anisotropic properties and sometimes are almost perfectly one-dimensional.

Historical Background

By 1928, the idea of a molecular orbital had been advanced by Robert Mulliken, who was influenced considerably by the work of Friedrich Hund. The LCAO method for approximating molecular orbitals was introduced in 1928 by B. N. Finklestein and G. E. Horowitz, while the LCAO method for solids was developed by Felix Bloch, as part of his doctoral dissertation in 1928, concurrently with and independent of the LCAO-MO approach. A much simpler interpolation scheme for approximating the electronic band structure, especially for the d-bands of transition metals, is the parameterized tight-binding method conceived in 1954 by John Clarke Slater and George Fred Koster, sometimes referred to as the SK tight-binding method. With the SK tight-binding method, electronic band structure calculations on a solid need not be carried out with full rigor as in the original Bloch›s theorem but, rather, first-principles calculations are carried out only at high-symmetry points and the band structure is interpolated over the remainder of the Brillouin zone between these points.

In this approach, interactions between different atomic sites are considered as perturbations. There exist several kinds of interactions we must consider. The crystal Hamiltonian is only approximately a sum of atomic Hamiltonians located at different sites and atomic wave functions overlap adjacent atomic sites in the crystal, and so are not accurate representations of the exact wave function. There are further explanations in the next section with some mathematical expressions.

In the recent research about strongly correlated material the tight binding approach is basic approximation because highly localized electrons like 3-d transition metal electrons sometimes display strongly correlated behaviors. In this case, the role of electron-electron interaction must be considered using the many-body physics description.

The tight-binding model is typically used for calculations of electronic band structure and band gaps in the static regime. However, in combination with other methods such as the random phase approximation (RPA) model, the dynamic response of systems may also be studied.

Mathematical Formulation

We introduce the atomic orbitals $\varphi_m(\boldsymbol{r})$, which are eigenfunctions of the Hamiltonian H_{at} of a single isolated atom. When the atom is placed in a crystal, this atomic wave function overlaps adjacent atomic sites, and so are not true eigenfunctions of the crystal Hamiltonian. The overlap is less when electrons are tightly bound, which is the source of the descriptor “tight-binding”. Any corrections to the atomic potential ΔU required to obtain the true Hamiltonian H of the system, are assumed small:

$$H(\boldsymbol{r}) = \sum_{\boldsymbol{R}_n} H_{\mathrm{at}}(\boldsymbol{r} - \boldsymbol{R}_n) + \Delta U(\boldsymbol{r})$$

A solution ψ_m to the time-independent single electron Schrödinger equation is then approximated as a linear combination of atomic orbitals $\varphi_m(\boldsymbol{r} - \boldsymbol{R}_n)$:

$$\psi_m(\boldsymbol{r}) = \sum_{\boldsymbol{R}_n} b_m(\boldsymbol{R}_n)\, \varphi_m(\boldsymbol{r} - \boldsymbol{R}_n),$$

where m refers to the m-th atomic energy level and R_n locates an atomic site in the crystal lattice.

Translational Symmetry and Normalization

The Bloch theorem states that the wave function in a crystal can change under translation only by a phase factor:

$$\psi(\boldsymbol{r} + \boldsymbol{R}_\ell) = e^{i\boldsymbol{k}\cdot\boldsymbol{R}_\ell}\psi(\boldsymbol{r}),$$

where k is the wave vector of the wave function. Consequently, the coefficients satisfy

$$\sum_{\boldsymbol{R}_n} b_m(\boldsymbol{R}_n)\, \varphi_m(\boldsymbol{r} - \boldsymbol{R}_n + \boldsymbol{R}_\ell) = e^{i\boldsymbol{k}\cdot\boldsymbol{R}_\ell} \sum_{\boldsymbol{R}_n} b_m(\boldsymbol{R}_n)\, \varphi_m(\boldsymbol{r} - \boldsymbol{R}_n).$$

By substituting $\boldsymbol{R}_p = \boldsymbol{R}_n - \boldsymbol{R}_\ell$, we find

$b_m(\boldsymbol{R}_p + \boldsymbol{R}_\ell) = e^{i\boldsymbol{k}\cdot\boldsymbol{R}_\ell} b_m(\boldsymbol{R}_p)$, (where in RHS we have replaced the dummy index $\boldsymbol{R}_n$ with $\boldsymbol{R}_p$)

or

$$b_m(\boldsymbol{R}_l) = e^{i\boldsymbol{k}\cdot\boldsymbol{R}_l} b_m(\boldsymbol{0}).$$

Normalizing the wave function to unity:

$$\int d^3r\, \psi_m^*(\boldsymbol{r})\psi_m(\boldsymbol{r}) = 1$$

$$= \sum_{\boldsymbol{R}_n} b_m^*(\boldsymbol{R}_n) \sum_{\boldsymbol{R}_\ell} b_m(\boldsymbol{R}_\ell) \int d^3r\, \varphi_m^*(\boldsymbol{r} - \boldsymbol{R}_n)\varphi_m(\boldsymbol{r} - \boldsymbol{R}_\ell)$$

$$= b_m^*(0)b_m(0) \sum_{\boldsymbol{R}_n} e^{-i\boldsymbol{k}\cdot\boldsymbol{R}_n} \sum_{\boldsymbol{R}_\ell} e^{i\boldsymbol{k}\cdot\boldsymbol{R}_\ell} \int d^3r\, \varphi_m^*(\boldsymbol{r} - \boldsymbol{R}_n)\varphi_m(\boldsymbol{r} - \boldsymbol{R}_\ell)$$

$$= N b_m^*(0)b_m(0) \sum_{\boldsymbol{R}_p} e^{-i\boldsymbol{k}\cdot\boldsymbol{R}_p} \int d^3r\, \varphi_m^*(\boldsymbol{r} - \boldsymbol{R}_p)\varphi_m(\boldsymbol{r})$$

$$= N b_m^*(0)b_m(0) \sum_{\boldsymbol{R}_p} e^{i\boldsymbol{k}\cdot\boldsymbol{R}_p} \int d^3r\, \varphi_m^*(\boldsymbol{r})\varphi_m(\boldsymbol{r} - \boldsymbol{R}_p),$$

so the normalization sets $b_m(0)$ as

$$b_m^*(0)b_m(0) = \frac{1}{N} \cdot \frac{1}{1 + \sum_{\boldsymbol{R}_p \neq 0} e^{i\boldsymbol{k}\cdot\boldsymbol{R}_p} \alpha_m(\boldsymbol{R}_p)},$$

where α_m ($\boldsymbol{R}_p$) are the atomic overlap integrals, which frequently are neglected resulting in

$$b_m(0) \approx \frac{1}{\sqrt{N}},$$

and

$$\psi_m(\boldsymbol{r}) \approx \frac{1}{\sqrt{N}} \sum_{\boldsymbol{R}_n} e^{i\boldsymbol{k}\cdot\boldsymbol{R}_n} \varphi_m(\boldsymbol{r}-\boldsymbol{R}_n).$$

The tight Binding Hamiltonian

Using the tight binding form for the wave function, and assuming only the m-th atomic energy level is important for the m-th energy band, the Bloch energies {\displaystyle \ varepsilon _{m}} are of the form

$$\begin{aligned}
\varepsilon_m &= \int d^3r\, \psi^*(\boldsymbol{r}) H(\boldsymbol{r}) \psi(\boldsymbol{r}) \\
&= \sum_{\boldsymbol{R}_n} b^*(\boldsymbol{R}_n) \int d^3r\, \varphi^*(\boldsymbol{r}-\boldsymbol{R}_n) H(\boldsymbol{r}) \psi(\boldsymbol{r}) \\
&= \sum_{\boldsymbol{R}_\ell} \sum_{\boldsymbol{R}_n} b^*(\boldsymbol{R}_n) \int d^3r\, \varphi^*(\boldsymbol{r}-\boldsymbol{R}_n) H_{\rm at}(\boldsymbol{r}-\boldsymbol{R}_\ell) \psi(\boldsymbol{r}) + \sum_{\boldsymbol{R}_n} b^*(\boldsymbol{R}_n) \int d^3r\, \varphi^*(\boldsymbol{r}-\boldsymbol{R}_n) \Delta U(\boldsymbol{r}) \psi(\boldsymbol{r}). \\
&\approx E_m + b^*(0) \sum_{\boldsymbol{R}_n} e^{-i\boldsymbol{k}\cdot\boldsymbol{R}_n} \int d^3r\, \varphi^*(\boldsymbol{r}-\boldsymbol{R}_n) \Delta U(\boldsymbol{r}) \psi(\boldsymbol{r}).
\end{aligned}$$

Here terms involving the atomic Hamiltonian at sites other than where it is centered are neglected. The energy then becomes

$$\begin{aligned}
\varepsilon_m(\boldsymbol{k}) &= E_m - N\,|b(0)|^2 \left(\beta_m + \sum_{\boldsymbol{R}_n \neq 0} \sum_l \gamma_{m,l}(\boldsymbol{R}_n) e^{i\boldsymbol{k}\cdot\boldsymbol{R}_n} \right), \\
&= E_m - \frac{\beta_m + \sum_{\boldsymbol{R}_n \neq 0} \sum_l e^{i\boldsymbol{k}\cdot\boldsymbol{R}_n} \gamma_{m,l}(\boldsymbol{R}_n)}{1 + \sum_{\boldsymbol{R}_n \neq 0} \sum_l e^{i\boldsymbol{k}\cdot\boldsymbol{R}_n} \alpha_{m,l}(\boldsymbol{R}_n)},
\end{aligned}$$

where E_m is the energy of the m-th atomic level, and $\alpha_{m,l}$, β_m and $\gamma_{m,l}$ are the tight binding matrix elements.

The Tight Binding Matrix Elements

The element

$$\beta_m = -\int \varphi_m^*(\boldsymbol{r}) \Delta U(\boldsymbol{r}) \varphi_m(\boldsymbol{r})\, d^3r,$$

is the atomic energy shift due to the potential on neighboring atoms. This term is relatively small in most cases. If it is large it means that potentials on neighboring atoms have a large influence on the energy of the central atom.

The next term

$$\gamma_{m,l}(\boldsymbol{R_n}) = -\int \varphi_m^*(\boldsymbol{r})\Delta U(\boldsymbol{r})\varphi_l(\boldsymbol{r}-\boldsymbol{R_n})\,d^3\boldsymbol{r}\,,$$

is the inter atomic matrix element between the atomic orbitals m and l on adjacent atoms. It is also called the bond energy or two center integral and it is the **most important element** in the tight binding model.

The last terms

$$\alpha_{m,l}(\boldsymbol{R_n}) = \int \varphi_m^*(\boldsymbol{r})\varphi_l(\boldsymbol{r}-\boldsymbol{R_n})\,d^3\boldsymbol{r}\,,$$

denote the overlap integrals between the atomic orbitals m and l on adjacent atoms.

Evaluation of the Matrix Elements

As mentioned before the values of the β_m-matrix elements are not so large in comparison with the ionization energy because the potentials of neighboring atoms on the central atom are limited. If β_m is not relatively small it means that the potential of the neighboring atom on the central atom is not small either. In that case it is an indication that the tight binding model is not a very good model for the description of the band structure for some reason. The inter atomic distances can be too small or the charges on the atoms or ions in the lattice is wrong for example.

The inter atomic matrix elements $\gamma_{m,l}$ can be calculated directly if the atomic wave functions and the potentials are known in detail. Most often this is not the case. There are numerous ways to get parameters for these matrix elements. Parameters can be obtained from chemical bond energy data. Energies and eigenstates on some high symmetry points in the Brillouin zone can be evaluated and values integrals in the matrix elements can be matched with band structure data from other sources.

The inter atomic overlap matrix elements $\alpha_{m,l}$ should be rather small or neglectable. If they are large it is again an indication that the tight binding model is of limited value for some purposes. Large overlap is an indication for too short inter atomic distance for example. In metals and transition metals the broad s-band or sp-band can be fitted better to an existing band structure calculation by the introduction of next-nearest-neighbor matrix elements and overlap integrals but fits like that don't yield a very useful model for the electronic wave function of a metal. Broad bands in dense materials are better described by a nearly free electron model.

The tight binding model works particularly well in cases where the band width is small and the electrons are strongly localized, like in the case of d-bands and f-bands. The model also gives good results in the case of open crystal structures, like diamond or silicon, where the number of neighbors is small. The model can easily be combined with a nearly free electron model in a hybrid NFE-TB model.

Connection to Wannier functions

Bloch wave functions describe the electronic states in a periodic crystal lattice. Bloch functions can be represented as a Fourier series

$$\psi_m(\mathbf{k},\mathbf{r}) = \frac{1}{\sqrt{N}} \sum_n a_m(\mathbf{R_n},\mathbf{r}) e^{i\mathbf{k}\cdot\mathbf{R_n}} ,$$

where $\boldsymbol{R}_n$ denotes an atomic site in a periodic crystal lattice, $\boldsymbol{k}$ is the wave vector of the Bloch wave, $\boldsymbol{r}$ is the electron position, m is the band index, and the sum is over all N atomic sites. The Bloch wave is an exact eigensolution for the wave function of an electron in a periodic crystal potential corresponding to an energy E_m $(\boldsymbol{k})$, and is spread over the entire crystal volume.

Using the Fourier transform analysis, a spatially localized wave function for the m-th energy band can be constructed from multiple Bloch waves:

$$a_m(\mathbf{R_n},\mathbf{r}) = \frac{1}{\sqrt{N}} \sum_{\mathbf{k}} e^{-i\mathbf{k}\cdot\mathbf{R_n}} \psi_m(\mathbf{k},\mathbf{r}) = \frac{1}{\sqrt{N}} \sum_{\mathbf{k}} e^{i\mathbf{k}\cdot(\mathbf{r}-\mathbf{R_n})} u_m(\mathbf{k},\mathbf{r}).$$

These real space wave functions $a_m(\mathbf{R_n},\mathbf{r})$ are called Wannier functions, and are fairly closely localized to the atomic site $\boldsymbol{R}_n$. Of course, if we have exact Wannier functions, the exact Bloch functions can be derived using the inverse Fourier transform.

However it is not easy to calculate directly either Bloch functions or Wannier functions. An approximate approach is necessary in the calculation of electronic structures of solids. If we consider the extreme case of isolated atoms, the Wannier function would become an isolated atomic orbital. That limit suggests the choice of an atomic wave function as an approximate form for the Wannier function, the so-called tight binding approximation.

Second Quantization

Modern explanations of electronic structure like t-J model and Hubbard model are based on tight binding model. Tight binding can be understood by working under a second quantization formalism.

Using the atomic orbital as a basis state, the second quantization Hamiltonian operator in the tight binding framework can be written as:

$$H = -t \sum_{\langle i,j \rangle,\sigma} (c^\dagger_{i,\sigma} c_{j,\sigma} + h.\,c.),$$

$c^\dagger_{i\sigma}, c_{j\sigma}$ - creation and annihilation operators

σ - spin polarization

t - hopping integral

$\langle i,j \rangle$ -nearest neighbor index

Here, hopping integral t corresponds to the transfer integral γ in tight binding model. Considering extreme cases of $t \rightarrow 0$, it is impossible for an electron to hop into neighboring sites. This case is the isolated atomic system. If the hopping term is turned on $(t > 0)$ electrons can stay in both sites lowering their kinetic energy.

In the strongly correlated electron system, it is necessary to consider the electron-electron interaction. This term can be written in

$$H_{ee} = \frac{1}{2}\sum_{n,m,\sigma}\langle n_1m_1, n_2m_2|\frac{e^2}{|r_1 - r_2|}|n_3m_3, n_4m_4\rangle c^{\dagger}_{n_1m_1\sigma_1}c^{\dagger}_{n_2m_2\sigma_2}c_{n_4m_4\sigma_2}c_{n_3m_3\sigma_1}$$

This interaction Hamiltonian includes direct Coulomb interaction energy and exchange interaction energy between electrons. There are several novel physics induced from this electron-electron interaction energy, such as metal-insulator transitions (MIT), high-temperature superconductivity, and several quantum phase transitions.

Example: One-dimensional s-band

Here the tight binding model is illustrated with a **s-band model** for a string of atoms with a single s-orbital in a straight line with spacing a and σ bonds between atomic sites.

To find approximate eigenstates of the Hamiltonian, we can use a linear combination of the atomic orbitals

$$|k\rangle = \frac{1}{\sqrt{N}}\sum_{n=1}^{N} e^{inka}|n\rangle$$

where N = total number of sites and k is a real parameter with $-\frac{\pi}{a} \leqq k \leqq \frac{\pi}{a}$. (This wave function is normalized to unity by the leading factor 1/√N provided overlap of atomic wave functions is ignored.) Assuming only nearest neighbor overlap, the only non-zero matrix elements of the Hamiltonian can be expressed as

$$\langle n|H|n\rangle = E_0 = E_i - U\,.$$

$$\langle n \pm 1|H|n\rangle = -\Delta$$

$$\langle n|n\rangle = 1\,; \quad \langle n \pm 1|n\rangle = S\,.$$

The energy E_i is the ionization energy corresponding to the chosen atomic orbital and U is the energy shift of the orbital as a result of the potential of neighboring atoms. The $\langle n \pm 1|H|n\rangle = -\Delta$ elements, which are the Slater and Koster interatomic matrix elements, are the bond energies $E_{i,j}$. In this one dimensional s-band model we only have σ-bonds between the s-orbitals with bond energy $E_{s,s} = V_{ss\sigma}$. The overlap between states on neighboring atoms is S. We can derive the energy of the state $|k\rangle$ using the above equation:

$$H|k\rangle = \frac{1}{\sqrt{N}} \sum_n e^{inka} H|n\rangle$$

$$\langle k|H|k\rangle = \frac{1}{N} \sum_{n,\,m} e^{i(n-m)ka} \langle m|H|n\rangle$$

$$= \frac{1}{N} \sum_n \langle n|H|n\rangle + \frac{1}{N} \sum_n \langle n-1|H|n\rangle e^{+ika} + \frac{1}{N} \sum_n \langle n+1|H|n\rangle e^{-ika}$$

$$= E_0 - 2\Delta \cos(ka) ,$$

where, for example,

$$\frac{1}{N} \sum_n \langle n|H|n\rangle = E_0 \frac{1}{N} \sum_n 1 = E_0 ,$$

and

$$\frac{1}{N} \sum_n \langle n-1|H|n\rangle e^{+ika} = -\Delta e^{ika} \frac{1}{N} \sum_n 1 = -\Delta e^{ika} .$$

$$\frac{1}{N} \sum_n \langle n-1|n\rangle e^{+ika} = S e^{ika} \frac{1}{N} \sum_n 1 = S e^{ika} .$$

Thus the energy of this state $|k\rangle$ can be represented in the familiar form of the energy dispersion:

$$E(k) = \frac{E_0 - 2\Delta \cos(ka)}{1 + 2S \cos(ka)} .$$

- For $k = 0$ the energy is $E = (E_0 - 2\Delta)/(1 + 2S)$ and the state consists of a sum of all atomic orbitals. This state can be viewed as a chain of bonding orbitals.
- For $k = \pi/(2a)$ the energy is $E = E_0$ and the state consists of a sum of atomic orbitals which are a factor $e^{i\pi/2}$ out of phase. This state can be viewed as a chain of non-bonding orbitals.
- Finally for $k = \pi/a$ the energy is $E = (E_0 + 2\Delta)/(1 - 2S)$ and the state consists of an alternating sum of atomic orbitals. This state can be viewed as a chain of anti-bonding orbitals.

This example is readily extended to three dimensions, for example, to a body-centered cubic or face-centered cubic lattice by introducing the nearest neighbor vector locations in place of simply n a. Likewise, the method can be extended to multiple bands using multiple different atomic orbitals at each site. The general formulation above shows how these extensions can be accomplished.

Table of Interatomic Matrix Elements

In 1954 J.C. Slater and G.F. Koster published, mainly for the calculation of transition metal d-bands, a table of interatomic matrix elements

$$E_{i,j}(\vec{\mathbf{r}}_{n,n'}) = \langle n, i|H|n', j\rangle$$

which, with a little patience and effort, can also be derived from the cubic harmonic orbitals straightforwardly. The table expresses the matrix elements as functions of LCAO two-centre bond integrals between two cubic harmonic orbitals, i and j, on adjacent atoms. The bond integrals are for example the $V_{ss\sigma}$, $V_{pp\pi}$ and $V_{dd\delta}$ for sigma, pi and delta bonds (Notice that these integrals should also depend on the distance between the atoms, i.e. are a function of (l,m,n), even though it is not explicitly stated every time.).

The interatomic vector is expressed as

$$\vec{\mathbf{r}}_{n,n'} = (r_x, r_y, r_z) = d(l, m, n)$$

where d is the distance between the atoms and l, m and n are the direction cosines to the neighboring atom.

$$\begin{aligned}
E_{s,s} &= V_{ss\sigma} \\
E_{s,x} &= lV_{sp\sigma} \\
E_{x,x} &= l^2 V_{pp\sigma} + (1-l^2)V_{pp\pi} \\
E_{x,y} &= lmV_{pp\sigma} - lmV_{pp\pi} \\
E_{x,z} &= lnV_{pp\sigma} - lnV_{pp\pi} \\
E_{s,xy} &= \sqrt{3}lmV_{sd\sigma} \\
E_{s,x^2-y^2} &= \frac{\sqrt{3}}{2}(l^2-m^2)V_{sd\sigma} \\
E_{s,3z^2-r^2} &= [n^2-(l^2+m^2)/2]V_{sd\sigma} \\
E_{x,xy} &= \sqrt{3}l^2mV_{pd\sigma} + m(1-2l^2)V_{pd\pi} \\
E_{x,yz} &= \sqrt{3}lmnV_{pd\sigma} - 2lmnV_{pd\pi} \\
E_{x,zx} &= \sqrt{3}l^2nV_{pd\sigma} + n(1-2l^2)V_{pd\pi} \\
E_{x,x^2-y^2} &= \frac{\sqrt{3}}{2}l(l^2-m^2)V_{pd\sigma} + l(1-l^2+m^2)V_{pd\pi} \\
E_{y,x^2-y^2} &= \frac{\sqrt{3}}{2}m(l^2-m^2)V_{pd\sigma} - m(1+l^2-m^2)V_{pd\pi} \\
E_{z,x^2-y^2} &= \frac{\sqrt{3}}{2}n(l^2-m^2)V_{pd\sigma} - n(l^2-m^2)V_{pd\pi} \\
E_{x,3z^2-r^2} &= l[n^2-(l^2+m^2)/2]V_{pd\sigma} - \sqrt{3}ln^2V_{pd\pi} \\
E_{y,3z^2-r^2} &= m[n^2-(l^2+m^2)/2]V_{pd\sigma} - \sqrt{3}mn^2V_{pd\pi} \\
E_{z,3z^2-r^2} &= n[n^2-(l^2+m^2)/2]V_{pd\sigma} + \sqrt{3}n(l^2+m^2)V_{pd\pi} \\
E_{xy,xy} &= 3l^2m^2V_{dd\sigma} + (l^2+m^2-4l^2m^2)V_{dd\pi} + (n^2+l^2m^2)V_{dd\delta} \\
E_{xy,yz} &= 3lm^2nV_{dd\sigma} + ln(1-4m^2)V_{dd\pi} + ln(m^2-1)V_{dd\delta} \\
E_{xy,zx} &= 3l^2mnV_{dd\sigma} + mn(1-4l^2)V_{dd\pi} + mn(l^2-1)V_{dd\delta}
\end{aligned}$$

$$E_{xy,x^2-y^2} = \frac{3}{2}lm(l^2 - m^2)V_{dd\sigma} + 2lm(m^2 - l^2)V_{dd\pi} + [lm(l^2 - m^2)/2]V_{dd\delta}$$
$$E_{yz,x^2-y^2} = \frac{3}{2}mn(l^2 - m^2)V_{dd\sigma} - mn[1 + 2(l^2 - m^2)]V_{dd\pi} + mn[1 + (l^2 - m^2)/2]V_{dd\delta}$$
$$E_{zx,x^2-y^2} = \frac{3}{2}nl(l^2 - m^2)V_{dd\sigma} + nl[1 - 2(l^2 - m^2)]V_{dd\pi} - nl[1 - (l^2 - m^2)/2]V_{dd\delta}$$
$$E_{xy,3z^2-r^2} = \sqrt{3}\left[lm(n^2 - (l^2 + m^2)/2)]V_{dd\sigma} - 2lmn^2V_{dd\pi} + [lm(1 + n^2)/2]V_{dd\delta}\right]$$
$$E_{yz,3z^2-r^2} = \sqrt{3}\left[mn(n^2 - (l^2 + m^2)/2)V_{dd\sigma} + mn(l^2 + m^2 - n^2)V_{dd\pi} - [mn(l^2 + m^2)/2]V_{dd\delta}\right]$$
$$E_{zx,3z^2-r^2} = \sqrt{3}\left[ln(n^2 - (l^2 + m^2)/2)V_{dd\sigma} + ln(l^2 + m^2 - n^2)V_{dd\pi} - [ln(l^2 + m^2)/2]V_{dd\delta}\right]$$
$$E_{x^2-y^2,x^2-y^2} = \frac{3}{4}(l^2 - m^2)^2V_{dd\sigma} + [l^2 + m^2 - (l^2 - m^2)^2]V_{dd\pi} + [n^2 + (l^2 - m^2)^2/4]V_{dd\delta}$$
$$E_{x^2-y^2,3z^2-r^2} = \sqrt{3}\left[(l^2 - m^2)[n^2 - (l^2 + m^2)/2]V_{dd\sigma}/2 + n^2(m^2 - l^2)V_{dd\pi} + [(1 + n^2)(l^2 - m^2)/4]V_{dd\delta}\right]$$
$$E_{3z^2-r^2,3z^2-r^2} = [n^2 - (l^2 + m^2)/2]^2V_{dd\sigma} + 3n^2(l^2 + m^2)V_{dd\pi} + \frac{3}{4}(l^2 + m^2)^2V_{dd\delta}$$

Not all interatomic matrix elements are listed explicitly. Matrix elements that are not listed in this table can be constructed by permutation of indices and cosine directions of other matrix elements in the table.

MUFFIN-TIN APPROXIMATION

The **muffin-tin approximation** is a shape approximation of the potential field in an atomistic environment. It is most commonly employed in quantum mechanical simulations of electronic band structure in solids. The approximation was proposed by John C. Slater. Augmented plane wave method is a method which uses muffin-tin approximation. It is a method to approximate the energy states of an electron in a crystal lattice. The basis approximation lies in the potential in which the potential is assumed to be spherically symmetric in the muffin-tin region and constant in the interstitial region. Wave functions (the augmented plane waves) are constructed by matching solutions of the Schrödinger equation within each sphere with plane-wave solutions in the interstitial region, and linear combinations of these wave functions are then determined by the variational method Many modern electronic structure methods employ the approximation. Among them are the augmented plane wave (APW) method, the linear muffin-tin orbital method (LMTO) and various Green›s function methods. One application is found in the variational theory developed by Korringa (1947) and by Kohn and Rostoker (1954) referred to as the KKR method. This method has been adapted to treat random materials as well, where it is called the KKR coherent potential approximation.

In its simplest form, non-overlapping spheres are centered on the atomic positions. Within these regions, the screened potential experienced by an electron is approximated to be spherically symmetric about the given nucleus. In the remaining interstitial region, the potential is approximated as a constant. Continuity of the potential between the atom-centered spheres and interstitial region is enforced.

In the interstitial region of constant potential, the single electron wave functions can be expanded in terms of plane waves. In the atom-centered regions, the wave functions can be expanded in terms of spherical harmonics and the eigenfunctions of a radial Schrödinger equation. Such use of functions other than plane waves as basis functions is termed the augmented plane-wave approach (of which there are many variations). It allows for an efficient representation of single-particle wave functions in the vicinity of the atomic cores where they can vary rapidly (and where plane waves would be a poor choice on convergence grounds in the absence of a pseudopotential).

DENSITY FUNCTIONAL THEORY

Density functional theory (**DFT**) is a computational quantum mechanical modelling method used in physics, chemistry and materials science to investigate the electronic structure (principally the ground state) of many-body systems, in particular atoms, molecules, and the condensed phases. Using this theory, the properties of a many-electron system can be determined by using functionals, i.e. functions of another function, which in this case is the spatially dependent electron density. Hence the name density functional theory comes from the use of functionals of the electron density. DFT is among the most popular and versatile methods available in condensed-matter physics, computational physics, and computational chemistry.

DFT has been very popular for calculations in solid-state physics since the 1970s. However, DFT was not considered accurate enough for calculations in quantum chemistry until the 1990s, when the approximations used in the theory were greatly refined to better model the exchange and correlation interactions. Computational costs are relatively low when compared to traditional methods, such as exchange only Hartree–Fock theory and its descendants that include electron correlation.

Despite recent improvements, there are still difficulties in using density functional theory to properly describe intermolecular interactions (of critical importance to understanding chemical reactions), especially van der Waals forces (dispersion); charge transfer excitations; transition states, global potential energy surfaces, dopant interactions and some other strongly correlated systems; and in calculations of the band gap and ferromagnetism in semiconductors. Its incomplete treatment of dispersion can adversely affect the accuracy of DFT (at least when used alone and uncorrected) in the treatment of systems which are dominated by dispersion (e.g. interacting noble gas atoms) or where dispersion competes significantly with other effects (e.g. in biomolecules). The development of new DFT methods designed to overcome this problem, by alterations to the functional or by the inclusion of additive terms, is a current research topic.

Overview of Method

Although density functional theory has its conceptual roots in the Thomas–Fermi model, DFT was put on a firm theoretical footing by the two **Hohenberg–Kohn theorems** (H–K). The original H–K theorems held only for non-degenerate ground states in the absence of a magnetic field, although they have since been generalized to encompass these.

The first H–K theorem demonstrates that the ground state properties of a many-electron system are uniquely determined by an electron density that depends on only 3 spatial coordinates. It lays the groundwork for reducing the many-body problem of N electrons with 3N spatial coordinates to 3 spatial coordinates, through the use of functionals of the electron density. This theorem can be extended to the time-dependent domain to develop time-dependent density functional theory (TDDFT), which can be used to describe excited states.

The second H–K theorem defines an energy functional for the system and proves that the correct ground state electron density minimizes this energy functional.

Within the framework of Kohn–Sham DFT (KS DFT), the intractable many-body problem of interacting electrons in a static external potential is reduced to a tractable problem of non-interacting electrons moving in an effective potential. The effective potential includes the external potential and the effects of the Coulomb interactions between the electrons, e.g., the exchange and correlation interactions. Modeling the latter two interactions becomes the difficulty within KS DFT. The simplest approximation is the local-density approximation (LDA), which is based upon exact exchange energy for a uniform electron gas, which can be obtained from the Thomas–Fermi model, and from fits to the correlation energy for a uniform electron gas. Non-interacting systems are relatively easy to solve as the wavefunction can be represented as a Slater determinant of orbitals. Further, the kinetic energy functional of such a system is known exactly. The exchange-correlation part of the total-energy functional remains unknown and must be approximated.

Another approach, less popular than KS DFT but arguably more closely related to the spirit of the original H-K theorems, is orbital-free density functional theory (OFDFT), in which approximate functionals are also used for the kinetic energy of the non-interacting system.

Derivation and Formalism

As usual in many-body electronic structure calculations, the nuclei of the treated molecules or clusters are seen as fixed (the Born–Oppenheimer approximation), generating a static external potential V in which the electrons are moving. A stationary electronic state is then described by a wavefunction $\Psi(\vec{r}_1,\ldots,\vec{r}_N)$ satisfying the many-electron time-independent Schrödinger equation

$$\hat{H}\Psi = \left[\hat{T}+\hat{V}+\hat{U}\right]\Psi = \left[\sum_i^N\left(-\frac{\hbar^2}{2m_i}\nabla_i^2\right)+\sum_i^N V(\vec{r}_i)+\sum_{i<j}^N U(\vec{r}_i,\vec{r}_j)\right]\Psi = E\Psi$$

where, for the N-electron system, $\hat{H}$ is the Hamiltonian, E is the total energy, $\hat{T}$ is the kinetic energy, $\hat{V}$ is the kinetic energy, $\hat{U}$ is the potential energy from the external field due to positively charged nuclei, and $\hat{U}$ is the electron-electron interaction energy. The operators $\hat{T}$ and $\hat{U}$ are called universal operators as they are the same for any N-electron system, while V is system dependent. This complicated many-particle equation is not separable into simpler single-particle equations because of the interaction term $\hat{U}$.

There are many sophisticated methods for solving the many-body Schrödinger equation based on the expansion of the wavefunction in Slater determinants. While the simplest one is the Hartree–Fock method, more sophisticated approaches are usually categorized as post-Hartree–Fock methods. However, the problem with these methods is the huge computational effort, which makes it virtually impossible to apply them efficiently to larger, more complex systems.

Here DFT provides an appealing alternative, being much more versatile as it provides a way to systematically map the many-body problem, with $\hat{U}$, onto a single-body problem without $\hat{U}$. In DFT the key variable is the electron density $n(\vec{r})$ which for a normalized

Ψ is given by

$$n(\vec{r}) = N \int d^3 r_2 \cdots \int d^3 r_N \Psi^*(\vec{r}, \vec{r}_2, \ldots, \vec{r}_N) \Psi(\vec{r}, \vec{r}_2, \ldots, \vec{r}_N).$$

This relation can be reversed, i.e., for a given ground-state density $n_0(\vec{r})$ it is possible, in principle, to calculate the corresponding ground-state wavefunction $\Psi_0(\vec{r}_1, \ldots, \vec{r}_N)$. In other words, Ψ is a unique functional of n_0,

$$\Psi_0 = \Psi[n_0]$$

and consequently the ground-state expectation value of an observable $\hat{O}$ is also a functional of n_0

$$O[n_0] = \left\langle \Psi[n_0] \middle| \hat{O} \middle| \Psi[n_0] \right\rangle.$$

In particular, the ground-state energy is a functional of n_0

$$E_0 = E[n_0] = \left\langle \Psi[n_0] \middle| \hat{T} + \hat{V} + \hat{U} \middle| \Psi[n_0] \right\rangle$$

where the contribution of the external potential $\left\langle \Psi[n_0] \middle| \hat{V} \middle| \Psi[n_0] \right\rangle$ can be written explicitly in terms of the ground-state density n_0

$$V[n_0] = \int V(\vec{r}) n_0(\vec{r}) d^3 r.$$

More generally, the contribution of the external potential $\left\langle \Psi \middle| \hat{V} \middle| \Psi \right\rangle$ can be written explicitly in terms of the density n,

$$V[n] = \int V(\vec{r}) n(\vec{r}) d^3 r.$$

The functionals $T[n]$ and $U[n]$ are called universal functionals, while $V[n]$ is called a non-universal functional, as it depends on the system under study. Having specified a system, i.e., having specified $\hat{V}$, one then has to minimize the functional

$$E[n] = T[n] + U[n] + \int V(\vec{r}) n(\vec{r}) d^3 r$$

with respect to $n(\vec{r})$, assuming one has got reliable expressions for $T[n]$ and $U[n]$. A successful minimization of the energy functional will yield the ground-state density n_0 and thus all other ground-state observables.

The variational problems of minimizing the energy functional $E[n]$ can be solved by applying the Lagrangian method of undetermined multipliers. First, one considers an energy functional that doesn't explicitly have an electron-electron interaction energy term,

$$E_s[n] = \left\langle \Psi_s[n] \middle| \hat{T} + \hat{V}_s \middle| \Psi_s[n] \right\rangle$$

where $\hat{T}$ denotes the kinetic energy operator and $\hat{V}_s$ is an external effective potential in which the particles are moving, so that $n_s(\vec{r}) \stackrel{\text{def}}{=} n(\vec{r})$.

Thus, one can solve the so-called Kohn–Sham equations of this auxiliary non-interacting system,

$$\left[-\frac{\hbar^2}{2m}\nabla^2 + V_s(\vec{r})\right]\phi_i(\vec{r}) = \epsilon_i\phi_i(\vec{r})$$

which yields the orbitals ϕ_i that reproduce the density $n(\vec{r})$ of the original many-body system

$$n(\vec{r}) \stackrel{\text{def}}{=} n_s(\vec{r}) = \sum_i^N |\phi_i(\vec{r})|^2.$$

The effective single-particle potential can be written in more detail as

$$V_s(\vec{r}) = V(\vec{r}) + \int \frac{e^2 n_s(\vec{r}\,')}{|\vec{r} - \vec{r}\,'|} \mathrm{d}^3r' + V_{\mathrm{XC}}[n_s(\vec{r})]$$

where the second term denotes the so-called Hartree term describing the electron-electron Coulomb repulsion, while the last term V_{XC} is called the exchange-correlation potential. Here, V_{XC} includes all the many-particle interactions. Since the Hartree term and V_{XC} depend on $n(\vec{r})$, which depends on the ϕ_i, which in turn depend on V_s, the problem of solving the Kohn–Sham equation has to be done in a self-consistent (i.e., iterative) way. Usually one starts with an initial guess for V_s and solves the Kohn–Sham equations for the ϕ_i. From these one calculates a new density and starts again. This procedure is then repeated until convergence is reached. A non-iterative approximate formulation called Harris functional DFT is an alternative approach to this.

NOTE1: The one-to-one correspondence between electron density and single-particle potential is not so smooth. It contains kinds of non-analytic structure. $E_s[n]$ contains kinds of singularities, cuts and branches. This may indicate a limitation of our hope for representing exchange-correlation functional in a simple analytic form.

NOTE2: It is possible to extend the DFT idea to the case of Green function G instead of the density n. It is called as Luttinger–Ward functional (or kinds of similar functionals), written as $E[G]$. However, G is determined not as its minimum, but as its extremum. Thus we may have some theoretical and practical difficulties.

NOTE3: There is no one-to-one correspondence between one-body density matrix $n(\vec{r},\vec{r}\,')$ and the one-body potential $V(\vec{r},\vec{r}\,')$. (Remember that all the eigenvalues of $n(\vec{r},\vec{r}\,')$ is unity). In other words, it ends up with a theory similar as the Hartree-Fock (or hybrid) theory.

Relativistic Density Functional Theory (explicit Functional Forms)

The same theorems can be proven in the case of relativistic electrons thereby providing generalization of DFT for relativistic case. Unlike nonrelativistic theory in relativistic case it's possible to derive a few exact and explicit formulas for relativistic density functional.

Let's consider an electron in hydrogen-like ion obeying relativistic Dirac equation. Hamiltonian H for relativistic electron moving in the Coulomb potential can be chosen in the following form (atomic units are used):

$$H = c(\vec{\alpha}\cdot\vec{p}) + eV + mc^2\beta,$$

where $V = -\frac{eZ}{r}$ is Coulomb potential of a point-like nucleus, $\vec{p}$ is a momentum operator of electron, e, m and c are electron electric charge, mass and speed of light in vacuum constants respectively, and finally $\vec{\alpha}$ and β are set of Dirac 4×4 matrixes:

$$\vec{\alpha} = \begin{pmatrix} 0 & \vec{\sigma} \\ \vec{\sigma} & 0 \end{pmatrix}, \beta = \begin{pmatrix} I & 0 \\ 0 & -I \end{pmatrix}$$

To find out eigen functions and corresponding energies one solves the eigen function equation:

$$H\Psi = E\Psi,$$

where $\Psi = (\Psi(1), \Psi(2), \Psi(3), \Psi(4))^T$ is a four component wave function and E is associated eigen energy. It is demonstrated in the article that application of the virial theorem to eigen function equation produces the following formula for eigen energy of any bound state:

$$E = mc^2 \langle \Psi |\beta| \Psi \rangle = mc^2 \int |\Psi(1)|^2 + |\Psi(2)|^2 - |\Psi(3)|^2 - |\Psi(4)|^2 d\tau.$$

and analogously the virial theorem applied to the eigen function equation with squared Hamiltonian (see also references therein) yields:

$$E^2 = m^2c^4 + emc^2 \langle \Psi |V\beta| \Psi \rangle.$$

It's easy to see that both written above formulas represent density functionals. The former formula can be easily generalized for multi-electron case.

Approximations (exchange-Correlation Functionals)

The major problem with DFT is that the exact functionals for exchange and correlation are not known except for the free electron gas. However, approximations exist which permit the calculation of certain physical quantities quite accurately. In physics the most widely used approximation is the local-density approximation (LDA), where the functional depends only on the density at the coordinate where the functional is evaluated:

$$E_{\rm XC}^{\rm LDA}[n] = \int \epsilon_{\rm XC}(n) n(\vec{r}) {\rm d}^3 r.$$

The local spin-density approximation (LSDA) is a straightforward generalization of the LDA to include electron spin:

$$E_{\rm XC}^{\rm LSDA}[n_\uparrow, n_\downarrow] = \int \epsilon_{\rm XC}(n_\uparrow, n_\downarrow) n(\vec{r}) {\rm d}^3 r.$$

Highly accurate formulae for the exchange-correlation energy density $\epsilon_{\rm XC}(n_\uparrow, n_\downarrow)$ have been constructed from quantum Monte Carlo simulations of jellium.

The LDA assumes that the density is the same everywhere. Because of this, the LDA has a tendency to over-estimate the exchange-correlation energy. To correct for this tendency, it is common to expand in terms of the gradient of the density in order to account for the non-homogeneity of the true electron density. This allows for corrections based on the changes in density away from the coordinate. These expansions are referred to as generalized gradient approximations (GGA)and have the following form:

$$E_{XC}^{\rm GGA}[n_\uparrow, n_\downarrow] = \int \epsilon_{XC}(n_\uparrow, n_\downarrow, \vec{\nabla} n_\uparrow, \vec{\nabla} n_\downarrow) n(\vec{r}) {\rm d}^3 r.$$

Using the latter (GGA), very good results for molecular geometries and ground-state energies have been achieved.

Potentially more accurate than the GGA functionals are the meta-GGA functionals, a natural development after the GGA (generalized gradient approximation). Meta-GGA DFT functional in its original form includes the second derivative of the electron density (the Laplacian) whereas GGA includes only the density and its first derivative in the exchange-correlation potential.

Functionals of this type are, for example, TPSS and the Minnesota Functionals. These functionals include a further term in the expansion, depending on the density, the gradient of the density and the Laplacian (second derivative) of the density.

Difficulties in expressing the exchange part of the energy can be relieved by including a component of the exact exchange energy calculated from Hartree–Fock theory. Functionals of this type are known as hybrid functionals.

Generalizations to Include Magnetic Fields

The DFT formalism described above breaks down, to various degrees, in the presence of a vector potential, i.e. a magnetic field. In such a situation, the one-to-one mapping between the ground-state electron density and wavefunction is lost. Generalizations to include the effects of magnetic fields have led to two different theories: current density functional theory (CDFT) and magnetic field density functional theory (BDFT). In both these theories, the functional used for the exchange and correlation must be generalized to include more than just the electron density. In current density functional theory, developed by Vignale and Rasolt, the functionals become dependent on both the electron density and the paramagnetic current density. In magnetic field density functional theory, developed by Salsbury, Grayce and Harris, the functionals depend on the electron density and the magnetic field, and the functional form can depend on the form of the magnetic field. In both of these theories it has been difficult to develop functionals beyond their equivalent to LDA, which are also readily implementable computationally. Recently an extension by Pan and Sahni extended the Hohenberg-Kohn theorem for non constant magnetic fields using the density and the current density as fundamental variables.

Applications

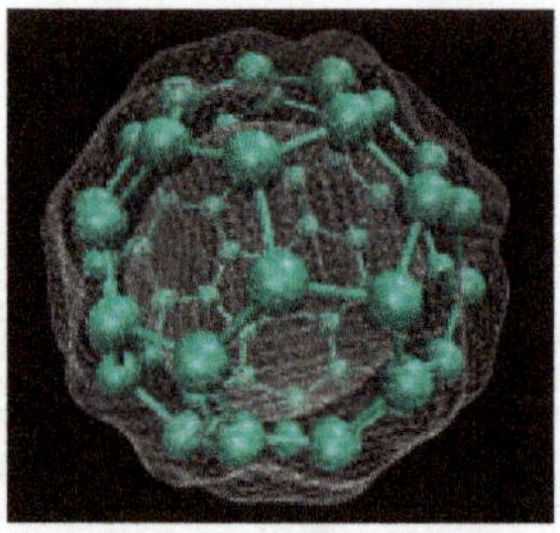

C_{60} with isosurface of ground-state electron density as calculated with DFT.

In general, density functional theory finds increasingly broad application in the chemical and material sciences for the interpretation and prediction of complex system behavior at an atomic scale. Specifically, DFT computational methods are applied for the study of systems to synthesis and processing parameters. In such systems, experimental studies are often encumbered by inconsistent results and non-equilibrium conditions. Examples of contemporary DFT applications include studying the effects of dopants on phase transformation behavior in oxides, magnetic behaviour in dilute magnetic semiconductor materials and the study of magnetic and electronic behavior in ferroelectrics and dilute magnetic semiconductors. Also, it has been shown that DFT has a good results in the prediction of sensitivity of some nanostructures to environment pollutants like SO_2 or Acrolein as well as prediction of mechanical properties.

In practice, Kohn–Sham theory can be applied in several distinct ways depending on what is being investigated. In solid state calculations, the local density approximations are still commonly used along with plane wave basis sets, as an electron gas approach is more appropriate for electrons delocalised through an infinite solid. In molecular calculations, however, more sophisticated functionals are needed, and a huge variety of exchange-correlation functionals have been developed for chemical applications. Some of these are inconsistent with the uniform electron gas approximation, however, they must reduce to LDA in the electron gas limit. Among physicists, probably the most widely used functional is the revised Perdew–Burke–Ernzerhof exchange model (a direct generalized-gradient parametrization of the free electron gas with no free parameters); however, this is not sufficiently calorimetrically accurate for gas-phase molecular calculations. In the chemistry community, one popular functional is known as BLYP (from the name Becke for the exchange part and Lee, Yang and Parr for the correlation part). Even more widely used is B3LYP which is a hybrid functional in which the exchange energy, in this case from Becke's exchange functional, is combined with the exact energy from Hartree–Fock theory. Along with the component exchange and correlation functionals, three parameters define the hybrid functional, specifying how much of the exact exchange is mixed in. The adjustable parameters in hybrid functionals are generally fitted to a 'training set' of molecules. Unfortunately, although the results obtained with these functionals are usually sufficiently accurate for most applications, there is no systematic way of improving them (in contrast to some of the traditional wavefunction-based methods like configuration interaction or coupled cluster theory). Hence in the current DFT approach it is not possible to estimate the error of the calculations without comparing them to other methods or experiments.

Thomas–Fermi model

The predecessor to density functional theory was the **Thomas–Fermi model**, developed independently by both Thomas and Fermi in 1927. They used a statistical model to approximate the distribution of electrons in an atom. The mathematical basis postulated that electrons are distributed uniformly in phase space with two electrons in every h^3 of volume. For each element of coordinate space volume d^3r we can fill out a sphere of momentum space up to the Fermi momentum p_f.

$$\frac{4}{3}\pi p_f^3(\vec{r}).$$

Equating the number of electrons in coordinate space to that in phase space gives:

$$n(\vec{r}) = \frac{8\pi}{3h^3} p_f^3(\vec{r}).$$

Solving for f and substituting into the classical kinetic energy formula then leads directly to a kinetic energy represented as a functional of the electron density:

$$t_{TF}[n] = \frac{p^2}{2m_e} \propto \frac{(n^{\frac{1}{3}})^2}{2m_e} \propto n^{\frac{2}{3}}(\vec{r})$$

$$T_{TF}[n] = C_F \int n(\vec{r}) n^{\frac{2}{3}}(\vec{r}) d^3r = C_F \int n^{\frac{5}{3}}(\vec{r}) d^3r$$

where $C_F = \frac{3h^2}{10m_e}\left(\frac{3}{8\pi}\right)^{\frac{2}{3}}.$

As such, they were able to calculate the energy of an atom using this kinetic energy functional combined with the classical expressions for the nuclear-electron and electron-electron interactions (which can both also be represented in terms of the electron density).

Although this was an important first step, the Thomas–Fermi equation›s accuracy is limited because the resulting kinetic energy functional is only approximate, and because the method does not attempt to represent the exchange energy of an atom as a conclusion of the Pauli principle. An exchange energy functional was added by Dirac in 1928.

However, the Thomas–Fermi–Dirac theory remained rather inaccurate for most applications. The largest source of error was in the representation of the kinetic energy, followed by the errors in the exchange energy, and due to the complete neglect of electron correlation.

Teller (1962) showed that Thomas–Fermi theory cannot describe molecular bonding. This can be overcome by improving the kinetic energy functional.

The kinetic energy functional can be improved by adding the Weizsäcker (1935) correction:

$$T_W[n] = \frac{\hbar^2}{8m} \int \frac{|\nabla n(\vec{r})|^2}{n(\vec{r})} d^3r.$$

Hohenberg–Kohn theorems

The Hohenberg-Kohn theorems relate to any system consisting of electrons moving under the influence of an external potential.

If two systems of electrons, one trapped in a potential $v_1(\vec{r})$ and the other in $v_2(\vec{r})$, have the same ground-state density $n(\vec{r})$ then necessarily $v_1(\vec{r}) - v_2(\vec{r}) = const$.

Corollary: the ground state density uniquely determines the potential and thus all properties of the system, including the many-body wave function. In particular, the "HK" functional, defined as $F[n] = T[n] + U[n]$ is a universal functional of the density (not depending explicitly on the external potential).

Theorem 2. The functional that delivers the ground state energy of the system, gives the lowest energy if and only if the input density is the true ground state density.

For any positive integer N and potential $v(\vec{r})$, a density functional $F[n]$ exists such that $E_{(v,N)}[n] = F[n] + \int v(\vec{r})n(\vec{r})d^3r$ obtains its minimal value at the ground-state density of N electrons in the potential $v(\vec{r})$. The minimal value of $E_{(v,N)}[n]$ is then the ground state energy of this system.

Pseudo-Potentials

The many electron Schrödinger equation can be very much simplified if electrons are divided in two groups: valence electrons and inner core electrons. The electrons in the inner shells are strongly bound and do not play a significant role in the chemical binding of atoms; they also partially screen the nucleus, thus forming with the nucleus an almost inert core. Binding properties are almost completely due to the valence electrons, especially in metals and semiconductors. This separation suggests that inner electrons can be ignored in a large number of cases, thereby reducing the atom to an ionic core that interacts with the valence electrons. The use of an effective interaction, a pseudopotential, that approximates the potential felt by the valence electrons, was first proposed by Fermi in 1934 and Hellmann in 1935. In spite of the simplification pseudo-potentials introduce in calculations, they remained forgotten until the late 50›s.

Ab initio Pseudo-potentials

A crucial step toward more realistic pseudo-potentials was given by Topp and Hopfield and more recently Cronin, who suggested that the pseudo-potential should be adjusted such that they describe the valence charge density accurately. Based on that idea, modern pseudo-potentials are obtained inverting the free atom Schrödinger equation for a given reference electronic configuration and forcing the pseudo wave-functions to coincide with the true valence wave functions beyond a certain distance $r l$. The pseudo wave-functions are also forced to have the same norm as the true valence wave-functions and can be written as

$$R_l^{PP}(r) = R_{nl}^{AE}(r).$$

$$\int_0^{rl} dr |R_l^{PP}(r)|^2 r^2 = \int_0^{rl} dr |R_{nl}^{AE}(r)|^2 r^2.$$

where $R_l(r)$. is the radial part of the wavefunction with angular momentum l., and pp. and AE. denote, respectively, the pseudo wave-function and the true (all-electron) wave-function. The index n in the true wave-functions denotes the valence level. The distance beyond which the true and the pseudo wave-functions are equal, rl., is also l-dependent.

Electron Smearing

The electrons of system will occupy the lowest Kohn-Sham eigenstates up to a given energy level according to the Aufbau principle. This corresponds to the step-like Fermi-Dirac distribution at absolute zero. If there are several degenerate or close to degenerate eigenstates at the Fermi level, it is possible to get convergence problems, since very small perturbations may change the electron occupation. One way of damping these oscillations is to smear the electrons, i.e. allowing fractional occupancies. One approach of doing this is to assign a finite temperature to the electron Fermi-Dirac distribution. Other ways is to assign a cumulative Gaussian distribution of the electrons or using a Methfessel-Paxton method.

K·P Perturbation Theory

In solid-state physics, **k·p perturbation theory** is an approximation scheme for calculating the band structure (particularly effective mass) and optical properties of crystalline solids. It is pronounced "k dot p", and is also called the "k·p method". This theory has been applied specifically in the framework of the **Luttinger–Kohn** model (after Joaquin Mazdak Luttinger and Walter Kohn), and of the **Kane model** (after Evan O. Kane).

Background and Derivation

Bloch's theorem and wavevectors

According to quantum mechanics (in the single-electron approximation), the electrons in any material have wavefunctions which can be described by the following Schrödinger equation:

$$\left(\frac{p^2}{2m} + V\right)\psi = E\psi$$

where **p** is the quantum-mechanical momentum operator, V is the potential, and m is the mass of an electron. (This equation neglects the spin-orbit effect; see below.)

In a crystalline solid, V is a periodic function, with the same periodicity as the crystal lattice. Bloch›s theorem proves that the solutions to this differential equation can be written as follows:

$$\psi_{n,\mathbf{k}}(\mathbf{x}) = e^{i\mathbf{k}\cdot\mathbf{x}}u_{n,\mathbf{k}}(\mathbf{x})$$

where **k** is a vector (called the wavevector), n is a discrete index (called the band index), and $u_{n,k}$ is a function with the same periodicity as the crystal lattice.

For any given n, the associated states are called a band. In each band, there will be a relation between the wavevector **k** and the energy of the state $E_{n,k}$, called the band dispersion. Calculating this dispersion is one of the primary applications of k·p perturbation theory.

Perturbation Theory

The periodic function $u_{n,k}$ satisfies the following Schrödinger-type equation:

$$H_{\mathbf{k}} u_{n,\mathbf{k}} = E_{n,\mathbf{k}} u_{n,\mathbf{k}}$$

where the Hamiltonian is

$$H_{\mathbf{k}} = \frac{p^2}{2m} + \frac{\hbar \mathbf{k} \cdot \mathbf{p}}{m} + \frac{\hbar^2 k^2}{2m} + V$$

Note that **k** is a vector consisting of three real numbers with dimensions of inverse length, while **p** is a vector of operators; to be explicit,

$$\mathbf{k} \cdot \mathbf{p} = k_x(-i\hbar\frac{\partial}{\partial x}) + k_y(-i\hbar\frac{\partial}{\partial y}) + k_z(-i\hbar\frac{\partial}{\partial z})$$

In any case, we write this Hamiltonian as the sum of two terms:

$$H = H_0 + H'_{\mathbf{k}}, \quad H_0 = \frac{p^2}{2m} + V, \quad H'_{\mathbf{k}} = \frac{\hbar^2 k^2}{2m} + \frac{\hbar \mathbf{k} \cdot \mathbf{p}}{m}$$

This expression is the basis for perturbation theory. The "unperturbed Hamiltonian" is H_0, which in fact equals the exact Hamiltonian at **k**=0 (i.e., at the Gamma point). The "perturbation" is the term $H'_{\mathbf{k}}$. The analysis that results is called "k·p perturbation theory", due to the term proportional to k·p. The result of this analysis is an expression for $E_{n,k}$ and $u_{n,k}$ in terms of the energies and wavefunctions at **k**=0.

Note that the "perturbation" term $H'_{\mathbf{k}}$ gets progressively smaller as **k** approaches zero. Therefore, k·p perturbation theory is most accurate for small values of **k**. However, if enough terms are included in the perturbative expansion, then the theory can in fact be reasonably accurate for any value of **k** in the entire Brillouin zone.

Expression for a nondegenerate band

For a nondegenerate band (i.e., a band which has a different energy at **k**=0 from any other band), with an extremum at **k**=0, and with no spin-orbit coupling, the result of k·p perturbation theory is (to lowest nontrivial order):

$$u_{n,\mathbf{k}} = u_{n,0} + \frac{\hbar}{m} \sum_{n' \neq n} \frac{\langle u_{n,0} | \mathbf{k} \cdot \mathbf{p} | u_{n',0} \rangle}{E_{n,0} - E_{n',0}} u_{n',0}$$

$$E_{n,\mathbf{k}} = E_{n,0} + \frac{\hbar^2 k^2}{2m} + \frac{\hbar^2}{m^2} \sum_{n' \neq n} \frac{|\langle u_{n,0} | \mathbf{k} \cdot \mathbf{p} | u_{n',0} \rangle|^2}{E_{n,0} - E_{n',0}}$$

Since **k** is a vector of real numbers (rather than a vector of more complicated linear operators), the matrix element in these expressions can be rewritten as:

$$\langle u_{n,0} | \mathbf{k} \cdot \mathbf{p} | u_{n',0} \rangle = \mathbf{k} \cdot \langle u_{n,0} | \mathbf{p} | u_{n',0} \rangle$$

Therefore you can calculate the energy at any **k** using only a few unknown parameters, namely $E_{n,0}$ and $\langle u_{n,0} | \mathbf{p} | u_{n',0} \rangle$. The latter are called "optical matrix elements", closely related to transition dipole moments. These parameters are typically inferred from experimental data.

In practice, the sum over n' often includes only the nearest one or two bands, since these tend to be the most important (due to the denominator). However, for improved accuracy, especially at larger **k**, more bands must be included, as well as more terms in the perturbative expansion than the ones written above.

Effective Mass

Using the expression above for the energy dispersion relation, a simplified expression for the effective mass in the conduction band of a semiconductor can be found. To approximate the dispersion relation in the case of the conduction band, take the energy E_{n0} as the minimum conduction band energy E_{c0} and include in the summation only terms with energies near the valence band maximum, where the energy difference in the denominator is smallest. (These terms are the largest contributions to the summation.) This denominator is then approximated as the band gap E_g, leading to an energy expression:

$$E_c(\boldsymbol{k}) \approx E_{c0} + \frac{(\hbar k)^2}{2m} + \frac{\hbar^2}{E_g m^2} \sum_n |\langle u_{c,0}|\mathbf{k}\cdot\mathbf{p}|u_{n,0}\rangle|^2$$

The effective mass in direction ℓ is then:

$$\frac{1}{m_\ell} = \frac{1}{\hbar^2} \sum_m \cdot \frac{\partial^2 E_c(\boldsymbol{k})}{\partial k_\ell \partial k_m} \approx \frac{1}{m} + \frac{2}{E_g m^2} \sum_{m,\,n} \langle u_{c,0}|p_\ell|u_{n,0}\rangle \langle u_{n,0}|p_m|u_{c,0}\rangle$$

Ignoring the details of the matrix elements, the key consequences are that the effective mass varies with the smallest bandgap and goes to zero as the gap goes to zero. A useful approximation for the matrix elements in direct gap semiconductors is:

$$\frac{2}{E_g m^2} \sum_{m,\,n} |\langle u_{c,0}|p_\ell|u_{n,0}\rangle||\langle u_{c,0}|p_m|u_{n,0}\rangle| \approx 20\text{eV}\frac{1}{mE_g},$$

which applies within about 15% or better to most group-IV, III-V and II-VI semiconductors.

In contrast to this simple approximation, in the case of valence band energy the spin-orbit interaction must be introduced (see below) and many more bands must be individually considered. The calculation is provided in Yu and Cardona. In the valence band the mobile carriers are holes. One finds there are two types of hole, named heavy and light, with anisotropic masses.

k·p model with spin-orbit interaction

Including the spin-orbit interaction, the Schrödinger equation for u is:

$$H_{\mathbf{k}} u_{n,\mathbf{k}} = E_{n,\mathbf{k}} u_{n,\mathbf{k}}$$

where[7]

$$H_{\mathbf{k}} = \frac{p^2}{2m} + \frac{\hbar}{m}\mathbf{k}\cdot\mathbf{p} + \frac{\hbar^2 k^2}{2m} + V + \frac{\hbar}{4m^2c^2}(\nabla V \times (\mathbf{p} + \hbar\mathbf{k}))\cdot\vec{\sigma}$$

where $\vec{\sigma} = (\sigma_x, \sigma_y, \sigma_z)$ is a vector consisting of the three Pauli matrices. This Hamiltonian can be subjected to the same sort of perturbation-theory analysis as above.

Calculation in Degenerate Case

For degenerate or nearly degenerate bands, in particular the valence bands in certain materials such as gallium arsenide, the equations can be analyzed by the methods of degenerate perturbation theory. Models of this type include the "Luttinger-Kohn model" (a.k.a. "Kohn-Luttinger model"), and the "Kane model".

EMPTY LATTICE APPROXIMATION

The **empty lattice approximation** is a theoretical electronic band structure model in which the potential is periodic and weak (close to constant). One may also consider an empty irregular lattice, in which the potential is not even periodic. The empty lattice approximation describes a number of properties of energy dispersion relations of non-interacting free electrons that move through a crystal lattice. The energy of the electrons in the "empty lattice" is the same as the energy of free electrons. The model is useful because it clearly illustrates a number of the sometimes very complex features of energy dispersion relations in solids which are fundamental to all electronic band structures.

Scattering and Periodicity

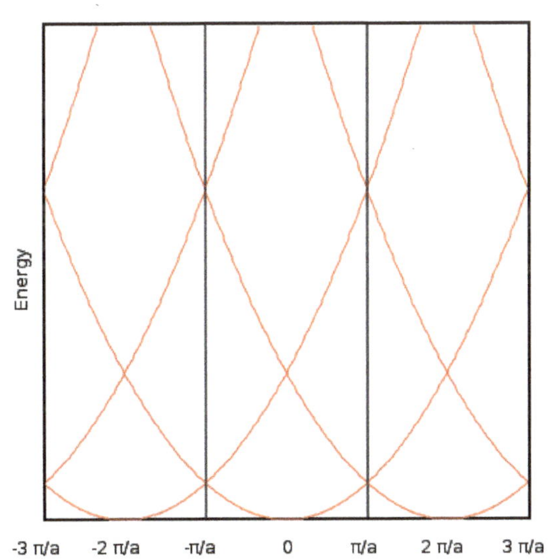

Free electron bands in a one dimensional lattice.

The periodic potential of the lattice in this free electron model must be weak because otherwise the electrons wouldn't be free. The strength of the scattering mainly depends on the geometry and topology of the system. Topologically defined parameters, like scattering cross sections, depend on the magnitude of the potential and the size of the potential well. For 1-, 2- and 3-dimensional spaces potential wells do always scatter waves, no matter how small their potentials are, what their signs are or how limited their sizes are. For a particle in a one-dimensional lattice, like the Kronig-Penney model, it is possible to calculate the band structure analytically by substituting the values for the potential, the lattice spacing and the size of potential well. For two and three-dimensional problems it is more difficult to calculate a band structure based on a similar model with a few parameters accurately. Nevertheless, the properties of the band structure can easily be approximated in most regions by perturbation methods.

In theory the lattice is infinitely large, so a weak periodic scattering potential will eventually be strong enough to reflect the wave. The scattering process results in the well known Bragg reflections of electrons by the periodic potential of the crystal structure. This is the origin of the periodicity of the dispersion relation and the division of k-space in Brillouin zones. The periodic energy dispersion relation is expressed as:

$$E_n(\mathbf{k}) = \frac{\hbar^2(\mathbf{k}+\mathbf{G}_n)^2}{2m}$$

The $\mathbf{G}_n$ are the reciprocal lattice vectors to which the bands $E_n(\mathbf{k})$ belong.

The figure on the right shows the dispersion relation for three periods in reciprocal space of a one-dimensional lattice with lattice cells of length a.

The Energy Bands and the Density of States

In a one-dimensional lattice the number of reciprocal lattice vectors $\mathbf{G}_n$ that determine the bands in an energy interval is limited to two when the energy rises. In two and three dimensional lattices the number of reciprocal lattice vectors that determine the free electron bands $E_n(\mathbf{k})$ increases more rapidly when the length of the wave vector increases and the energy rises. This is because the number of reciprocal lattice vectors $\mathbf{G}_n$ that lie in an interval $[\mathbf{k}, \mathbf{k}+d\mathbf{k}]$ increases. The density of states in an energy interval $[E, E+dE]$ depends on the number of states in an interval $[\mathbf{k}, \mathbf{k}+d\mathbf{k}]$ in reciprocal space and the slope of the dispersion relation $E_n(\mathbf{k})$.

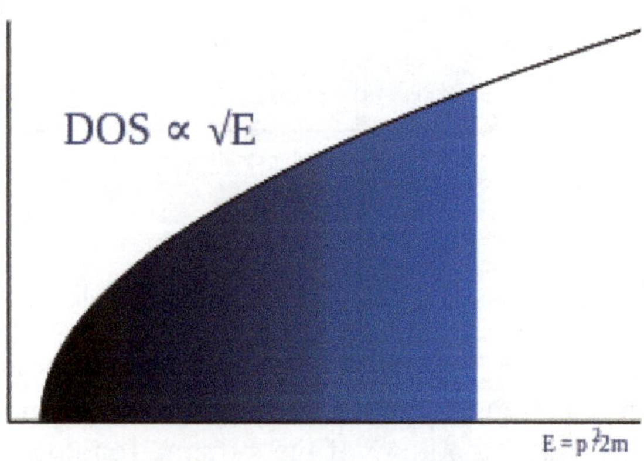

Figure 3: Free-electron DOS in 3-dimensional k-space.

Though the lattice cells are not spherically symmetric, the dispersion relation still has spherical symmetry from the point of view of a fixed central point in a reciprocal lattice cell if the dispersion relation is extended outside the central Brillouin zone. The density of states in a three-dimensional lattice will be the same as in the case of the absence of a lattice. For the three-dimensional case the density of states $D_3(E)$;

$$D_3(E) = 2\pi\sqrt{\frac{E-E_0}{c_k^3}}.$$

In three-dimensional space the Brillouin zone boundaries are planes. The dispersion relations show conics of the free-electron energy dispersion parabolas for all possible reciprocal lattice vectors. This results in a very complicated set intersecting of curves when the dispersion relations are calculated because there is a large number of possible angles between evaluation trajectories, first and higher order Brillouin zone boundaries and dispersion parabola intersection cones.

Second, Third and Higher Brillouin zones

"Free electrons" that move through the lattice of a solid with wave vectors {\displaystyle {\mathbf {k}}} far outside the first Brillouin zone are still reflected back into the first Brillouin zone. See the external links section for sites with examples and figures.

The Nearly Free Electron Model

In most simple metals, like aluminium, the screening effect strongly reduces the electric field of the ions in the solid. The electrostatic potential is expressed as

$$V(r) = \frac{Ze}{r}e^{-qr}$$

where Z is the atomic number, e is the elementary unit charge, r is the distance to the nucleus of the embedded ion and q is a screening parameter that determines the range of the potential. The Fourier transform, $U_{\mathbf{G}}$, of the lattice potential, $V(\mathbf{r})$, is expressed as

$$U_{\mathbf{G}} = \frac{4\pi Ze}{q^2 + \mathbf{G}^2}$$

When the values of the off-diagonal elements $U_{\mathbf{G}}$ between the reciprocal lattice vectors in the Hamiltonian almost go to zero. As a result, the magnitude of the band gap $2|U_{\mathbf{G}}|$ collapses and the empty lattice approximation is obtained.

The Electron Bands of Common Metal Crystals

Apart from a few exotic exceptions, metals crystallize in three kinds of crystal structures: the BCC and FCC cubic crystal structures and the hexagonal close-packed HCP crystal structure.

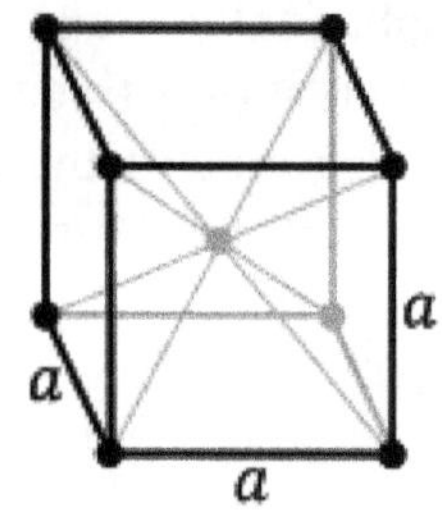

Body-centered cubic (I)

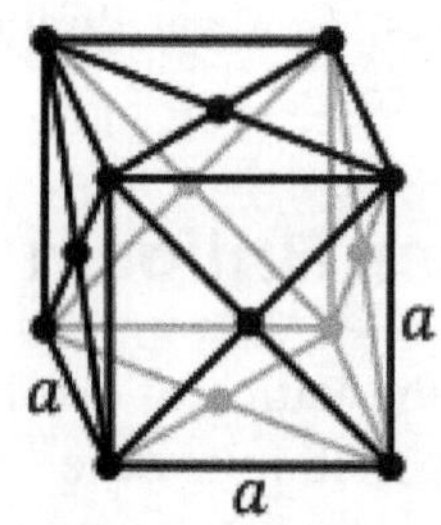

Face-centered cubic (F)

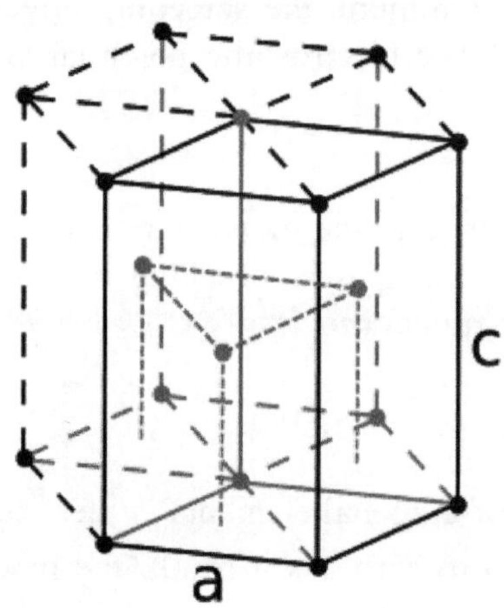

Hexagonal close-packed

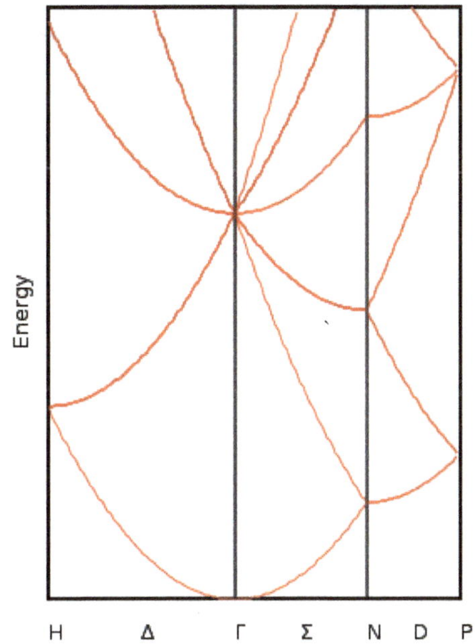

Free electron bands in a BCC crystal structure

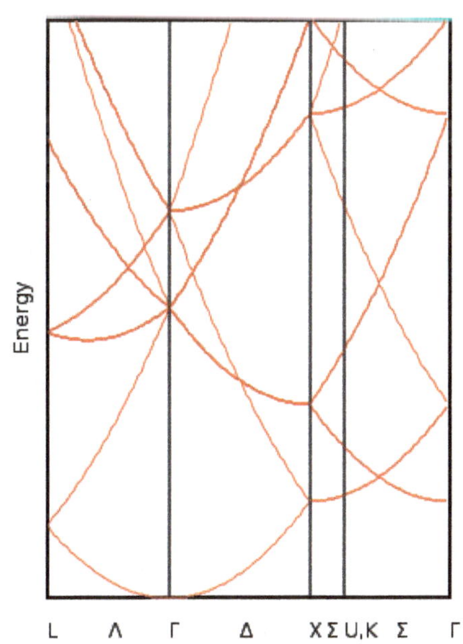

Free electron bands in a FCC crystal structure.

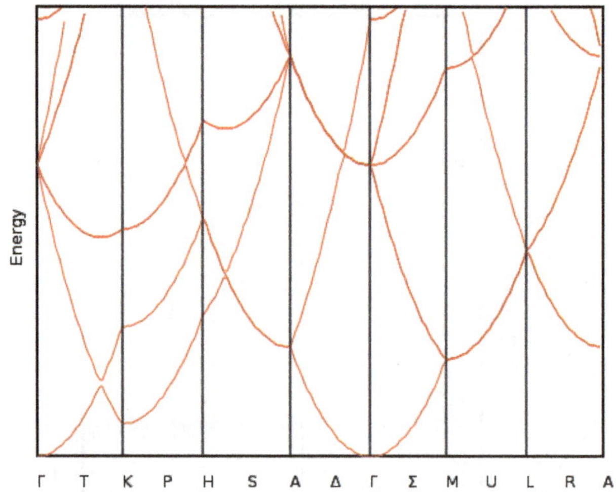

Free electron bands in a HCP crystal structure.

REFERNCES

1. Assadi, M.H.N; et al. (2013). "Theoretical study on copper›s energetics and magnetism in TiO_2 polymorphs". Journal of Applied Physics. 113 (23): 233913. arXiv:1304.1854. Bibcode:2013JAP...113w3913A. doi:10.1063/1.4811539.
2. Van Mourik, Tanja; Gdanitz, Robert J. (2002). "A critical note on density functional theory studies on rare-gas dimers". Journal of Chemical Physics. 116 (22): 9620–9623. Bibcode:2002JChPh.116.9620V. doi:10.1063/1.1476010.
3. Vondrášek, Jiří; Bendová, Lada; Klusák, Vojtěch; Hobza, Pavel (2005). "Unexpectedly strong energy stabilization inside the hydrophobic core of small protein rubredoxin mediated by aromatic residues: correlated ab initio quantum chemical calculations". Journal of the American Chemical Society. 127 (8): 2615–2619. doi:10.1021/ja044607h. PMID 15725017.
4. Grimme, Stefan (2006). "Semiempirical hybrid density functional with perturbative second-order correlation". Journal of Chemical Physics. 124 (3): 034108. Bibcode:2006JChPh.124c4108G. doi:10.1063/1.2148954. PMID 16438568.
5. Zimmerli, Urs; Parrinello, Michele; Koumoutsakos, Petros (2004). "Dispersion corrections to density functionals for water aromatic interactions". Journal of Chemical Physics. 120 (6): 2693–2699. Bibcode:2004JChPh.120.2693Z. doi:10.1063/1.1637034. PMID 15268413.
6. Grimme, Stefan (2004). "Accurate description of van der Waals complexes by density functional theory including empirical corrections". Journal of Computational Chemistry. 25 (12): 1463–1473. doi:10.1002/jcc.20078. PMID 15224390.

7. Von Lilienfeld, O. Anatole; Tavernelli, Ivano; Rothlisberger, Ursula; Sebastiani, Daniel (2004). "Optimization of effective atom centered potentials for London dispersion forces in density functional theory". Physical Review Letters. 93 (15): 153004. Bibcode:2004PhRvL..93o3004V. doi:10.1103/PhysRevLett.93.153004. PMID 15524874.

8. Tkatchenko, Alexandre; Scheffler, Matthias (2009). "Accurate Molecular Van Der Waals Interactions from Ground-State Electron Density and Free-Atom Reference Data". Physical Review Letters. 102 (7): 073005. Bibcode:2009PhRvL.102g3005T. doi:10.1103/PhysRevLett.102.073005. PMID 19257665.

9. Hohenberg, Pierre; Walter Kohn (1964). "Inhomogeneous electron gas". Physical Review. 136 (3B): B864–B871. Bibcode:1964PhRv..136..864H. doi:10.1103/PhysRev.136.B864.

10. Levy, Mel (1979). "Universal variational functionals of electron densities, first-order density matrices, and natural spin-orbitals and solution of the v-representability problem". Proceedings of the National Academy of Sciences. United States National Academy of Sciences. 76 (12): 6062–6065. Bibcode:1979PNAS...76.6062L. doi:10.1073/pnas.76.12.6062.

11. Vignale, G.; Mark Rasolt (1987). "Density-functional theory in strong magnetic fields". Physical Review Letters. American Physical Society. 59 (20): 2360–2363. Bibcode:1987PhRvL..59.2360V. doi:10.1103/PhysRevLett.59.2360. PMID 10035523.

12. Kohn, W.; Sham, L. J. (1965). "Self-consistent equations including exchange and correlation effects". Physical Review. 140 (4A): A1133–A1138. Bibcode:1965PhRv..140.1133K. doi:10.1103/PhysRev.140.A1133.

13. M. Brack (1983), "Virial theorems for relativistic spin-½ and spin-0 particles", Phys. Rev. D, 27: 1950, doi:10.1103/physrevd.27.1950

14. K. Koshelev (2015). "About density functional theory interpretation". arXiv:0812.2919 [quant-ph].

15. K. Koshelev (2007). "Alpha variation problem and q-factor definition". arXiv:0707.1146 [physics.atom-ph].

16. Kieron Burke; Lucas O. Wagner (2013). "DFT in a nutshell". International Journal of Quantum Chemistry. 113 (2): 96. doi:10.1002/qua.24259.

17. John P. Perdew; Adrienn Ruzsinszky; Jianmin Tao; Viktor N. Staroverov; Gustavo Scuseria; Gábor I. Csonka (2005). "Prescriptions for the design and selection of density functional approximations: More constraint satisfaction with fewer fits". Journal of Chemical Physics. 123 (6): 062201. Bibcode:2005JChPh.123f2201P. doi:10.1063/1.1904565. PMID 16122287.

18. Becke, Axel D. (2014-05-14). "Perspective: Fifty years of density-functional theory in chemical physics". The Journal of Chemical Physics. 140 (18): 18A301. Bibcode:2014JChPh.140rA301B. doi:10.1063/1.4869598. ISSN 0021-9606. PMID 24832308.

19. Perdew, John P; Chevary, J A; Vosko, S H; Jackson, Koblar, A; Pederson, Mark R; Singh, D J; Fiolhais, Carlos (1992). "Atoms, molecules, solids, and surfaces: Applications of the generalized gradient approximation for exchange and

correlation". Physical Review B. 46 (11): 6671. Bibcode:1992PhRvB..46.6671P. doi:10.1103/physrevb.46.6671.

20. Becke, Axel D (1988). "Density-functional exchange-energy approximation with correct asymptotic behavior". Physical Review A. 38 (6): 3098. Bibcode:1988PhRvA..38.3098B. doi:10.1103/physreva.38.3098. PMID 9900728.
21. Langreth, David C; Mehl, M J (1983). "Beyond the local-density approximation in calculations of ground-state electronic properties". Physical Review B. 28 (4): 1809. Bibcode:1983PhRvB..28.1809L. doi:10.1103/physrevb.28.1809.
22. Grayce, Christopher; Robert Harris (1994). "Magnetic-field density-functional theory". Physical Review A. 50 (4): 3089–3095. Bibcode:1994PhRvA..50.3089G. doi:10.1103/PhysRevA.50.3089. PMID 9911249.
23. Viraht, Xiao-Yin (2012). "Hohenberg-Kohn theorem including electron spin". Physical Review A. 86 (4): 042502. Bibcode:2012PhRvA..86d2502P. doi:10.1103/physreva.86.042502.
24. Segall, M.D.; Lindan, P.J (2002). "First-principles simulation: ideas, illustrations and the CASTEP code". Journal of Physics: Condensed Matter. 14 (11): 2717. Bibcode:2002JPCM...14.2717S. doi:10.1088/0953-8984/14/11/301.
25. Hanaor, Dorian A. H.; Assadi, Mohammed H. N.; Li, Sean; Yu, Aibing; Sorrell, Charles C. (2012). "Ab initio study of phase stability in doped TiO_2". Computational Mechanics. 50 (2): 185–194. doi:10.1007/s00466-012-0728-4.
26. Somayeh. F. Rastegar, Hamed Soleymanabadi (2014-01-01). "Theoretical investigation on the selective detection of SO2 molecule by AlN nanosheets". Journal of Molecular Modeling. 20 (9). doi:10.1007/s00894-014-2439-6.
27. Somayeh F. Rastegar, Hamed Soleymanabadi (2013-01-01). "DFT studies of acrolein molecule adsorption on pristine and Al- doped graphenes". Journal of Molecular Modeling. 19 (9): 3733–40. doi:10.1007/s00894-013-1898-5. PMID 23793719.
28. Music, D.; Geyer, R.W.; Schneider, J.M. (2016). "Recent progress and new directions in density functional theory based design of hard coatings". Surface & Coatings Technology. 286: 178. doi:10.1016/j.surfcoat.2015.12.021.
29. (Parr & Yang 1989, p. 47)
30. March, N. H. (1992). Electron Density Theory of Atoms and Molecules. Academic Press. p. 24. ISBN 0-12-470525-1.
31. Weizsäcker, C. F. v. (1935). "Zur Theorie der Kernmassen". Zeitschrift für Physik. 96 (7–8): 431–58. Bibcode:1935ZPhy...96..431W. doi:10.1007/BF01337700.
32. (Parr & Yang 1989, p. 127)
33. Topp, William C.; Hopfield, John J. (1973-02-15). "Chemically Motivated Pseudopotential for Sodium". Physical Review B. 7 (4): 1295–1303. Bibcode:1973PhRvB...7.1295T. doi:10.1103/PhysRevB.7.1295.
34. Michelini, M. C.; Pis Diez, R.; Jubert, A. H. (25 June 1998). "A Density Functional Study of Small Nickel Clusters". International Journal of Quantum

Chemistry. 70 (4–5): 694. doi:10.1002/(SICI)1097-461X(1998)70:4/5<693::AID-QUA15>3.0.CO;2-3. Retrieved 21 October 2016.

35. Finite temperature approaches -- smearing methods". VASP the GUIDE. Retrieved 21 October 2016.

36. Tong, Lianheng. "Methfessel-Paxton Approximation to Step Function". Metal CONQUEST. Retrieved 21 October 2016.

CHAPTER 9

Molecular Hamiltonian

Chapter Contents

- **Intoduction**
- **Coulomb Hamiltonian**
- **The Schrödinger Equation of the Coulomb Hamiltonian**
- **Clamped Nucleus Hamiltonian**
- **Watson's Nuclear Motion Hamiltonian**

In this discussion, we consider the formulation of the Schrodinger equation for diatomic molecules; this can be extended to larger molecules. First we will consider the separation of the total Hamiltonian for a 4-body problem into a more tractable form. We will afterward discuss the molecular wave functions. For the hydrogen molecule, we are concerned with 2 nuclei and 2 electrons. The total Hamiltonian, representing the total energy operator, is:

$$\begin{aligned}\hat{H}(\vec{r},\vec{R}) &= -\frac{\hbar^2}{2M}\left(\nabla_A^2+\nabla_B^2\right)-\frac{\hbar^2}{2m_e}\left(\nabla_1^2+\nabla_2^2\right)\\ &- \frac{Z_Ae^2}{4\pi\epsilon_0 r_{1A}}-\frac{Z_Ae^2}{4\pi\epsilon_0 r_{2A}}\\ &- \frac{Z_Be^2}{4\pi\epsilon_0 r_{1B}}-\frac{Z_Be^2}{4\pi\epsilon_0 r_{2B}}+\frac{e^2}{4\pi\epsilon_0 r_{12}}\\ &+ \frac{Z_AZ_Be^2}{4\pi\epsilon_0 R_{AB}}\end{aligned}$$

INTODUCTION

In atomic, molecular, and optical physics and quantum chemistry, the **molecular Hamiltonian** is the Hamiltonian operator representing the energy of the electrons and nuclei in a molecule. This operator and the associated Schrödinger equation play a central role in computational chemistry and physics for computing properties of molecules and aggregates of molecules, such as thermal conductivity, specific heat, electrical conductivity, optical, and magnetic properties, and reactivity.

The elementary parts of a molecule are the nuclei, characterized by their atomic numbers, Z, and the electrons, which have negative elementary charge, $-e$. Their interaction gives a nuclear charge of $Z + q$, where $q = -eN$, with N equal to the number of electrons. Electrons and nuclei are, to a very good approximation, point charges and point masses. The molecular Hamiltonian is a sum of several terms: its major terms are the kinetic energies of the electrons and the Coulomb (electrostatic) interactions between the two kinds of charged particles. The Hamiltonian that contains only the kinetic energies of electrons and nuclei, and the Coulomb interactions between them, is known as the **Coulomb Hamiltonian**. From it are missing a number of small terms, most of which are due to electronic and nuclear spin.

Although it is generally assumed that the solution of the time-independent Schrödinger equation associated with the Coulomb Hamiltonian will predict most properties of the molecule, including its shape (three-dimensional structure), calculations based on the full Coulomb Hamiltonian are very rare. The main reason is that its Schrödinger equation is very difficult to solve. Applications are restricted to small systems like the hydrogen molecule.

Almost all calculations of molecular wavefunctions are based on the separation of the Coulomb Hamiltonian first devised by Born and Oppenheimer. The nuclear kinetic energy terms are omitted from the Coulomb Hamiltonian and one considers the remaining Hamiltonian as a Hamiltonian of electrons only. The stationary nuclei enter the problem only as generators of an electric potential in which the electrons move in a quantum mechanical way. Within this framework the molecular Hamiltonian has been simplified to the so-called **clamped nucleus Hamiltonian**, also called **electronic Hamiltonian**, that acts only on functions of the electronic coordinates.

Once the Schrödinger equation of the clamped nucleus Hamiltonian has been solved for a sufficient number of constellations of the nuclei, an appropriate eigenvalue (usually the lowest) can be seen as a function of the nuclear coordinates, which leads to a potential energy surface. In practical calculations the surface is usually fitted in terms of some analytic functions. In the second step of the Born–Oppenheimer approximation the part of the full Coulomb Hamiltonian that depends on the electrons is replaced by the potential energy surface. This converts the total molecular Hamiltonian into another Hamiltonian that acts only on the nuclear coordinates. In the case of a breakdown of the Born–Oppenheimer approximation—which occurs when energies of different electronic states are close—the neighboring potential energy surfaces are needed, see this article for more details on this.

The nuclear motion Schrödinger equation can be solved in a space-fixed (laboratory) frame, but then the translational and rotational (external) energies are not accounted for. Only the (internal) atomic vibrations enter the problem. Further, for molecules larger than triatomic ones, it is quite common to introduce the harmonic approximation, which approximates the potential energy surface as a quadratic function of the atomic displacements. This gives the **harmonic nuclear motion Hamiltonian**. Making the harmonic approximation, we can convert the Hamiltonian into a sum of uncoupled one-dimensional harmonic oscillator Hamiltonians. The one-dimensional harmonic oscillator is one of the few systems that allows an exact solution of the Schrödinger equation.

Alternatively, the nuclear motion (rovibrational) Schrödinger equation can be solved in a special frame (an Eckart frame) that rotates and translates with the molecule. Formulated with respect to this body-fixed frame the Hamiltonian accounts for rotation, translation and vibration of the nuclei. Since Watson introduced in 1968 an important simplification to this Hamiltonian, it is often referred to as **Watson›s nuclear motion Hamiltonian**, but it is also known as the **Eckart Hamiltonian**.

COULOMB HAMILTONIAN

The algebraic form of many observables—i.e., Hermitian operators representing observable quantities—is obtained by the following quantization rules:

- Write the classical form of the observable in Hamilton form (as a function of momenta **p** and positions **q**). Both vectors are expressed with respect to an arbitrary inertial frame, usually referred to as *laboratory-frame* or *space-fixed frame*.
- Replace **p** by $-i\hbar\nabla$ and interpret **q** as a multiplicative operator. Here ∇ is the nabla operator, a vector operator consisting of first derivatives. The well-known commutation relations for the **p** and **q** operators follow directly from the differentiation rules.

Classically the electrons and nuclei in a molecule have kinetic energy of the form $p^2/(2\ m)$ and interact via Coulomb interactions, which are inversely proportional to the distance r_{ij} between particle i and j.

$$r_{ij} \equiv |\mathbf{r}_i - \mathbf{r}_j| = \sqrt{(\mathbf{r}_i - \mathbf{r}_j)\cdot(\mathbf{r}_i - \mathbf{r}_j)} = \sqrt{(x_i - x_j)^2 + (y_i - y_j)^2 + (z_i - z_j)^2}.$$

In this expression $\mathbf{r}_i$ stands for the coordinate vector of any particle (electron or nucleus), but from here on we will reserve capital **R** to represent the nuclear coordinate, and lower case **r** for the electrons of the system. The coordinates can be taken to be expressed with respect to any Cartesian frame centered anywhere in space, because distance, being an inner product, is invariant under rotation of the frame and, being the norm of a difference vector, distance is invariant under translation of the frame as well.

By quantizing the classical energy in Hamilton form one obtains the a molecular Hamilton operator that is often referred to as the **Coulomb Hamiltonian**. This Hamiltonian is a sum of five terms. They are

- The kinetic energy operators for each nucleus in the system;
- The kinetic energy operators for each electron in the system;

- The potential energy between the electrons and nuclei – the total electron-nucleus Coulombic attraction in the system;
- The potential energy arising from Coulombic electron-electron repulsions
- The potential energy arising from Coulombic nuclei-nuclei repulsions – also known as the nuclear repulsion energy. See electric potential for more details.

1. $\hat{T}_n = -\sum_i \frac{\hbar^2}{2M_i}\nabla^2_{\mathbf{R}_i}$

2. $\hat{T}_e = -\sum_i \frac{\hbar^2}{2m_e}\nabla^2_{\mathbf{r}_i}$

3. $\hat{U}_{en} = -\sum_i\sum_j \frac{Z_i e^2}{4\pi\epsilon_0 \left|\mathbf{R}_i - \mathbf{r}_j\right|}$

4. $\hat{U}_{ee} = \frac{1}{2}\sum_i\sum_{j\neq i} \frac{e^2}{4\pi\epsilon_0 \left|\mathbf{r}_i - \mathbf{r}_j\right|} = \sum_i\sum_{j>i} \frac{e^2}{4\pi\epsilon_0 \left|\mathbf{r}_i - \mathbf{r}_j\right|}$

5. $\hat{U}_{nn} = \frac{1}{2}\sum_i\sum_{j\neq i} \frac{Z_i Z_j e^2}{4\pi\epsilon_0 \left|\mathbf{R}_i - \mathbf{R}_j\right|} = \sum_i\sum_{j>i} \frac{Z_i Z_j e^2}{4\pi\epsilon_0 \left|\mathbf{R}_i - \mathbf{R}_j\right|}.$

Here M_i is the mass of nucleus i, Z_i is the atomic number of nucleus i, and m_e is the mass of the electron. The Laplace operator of particle i is $\nabla^2_{\mathbf{r}_i} \equiv \nabla_{\mathbf{r}_i}\cdot\nabla_{\mathbf{r}_i} = \frac{\partial^2}{\partial x_i^2} + \frac{\partial^2}{\partial y_i^2} + \frac{\partial^2}{\partial z_i^2}$. Since the kinetic energy operator is an inner product, it is invariant under rotation of the Cartesian frame with respect to which x_i, y_i, and z_i are expressed.

Small Terms

In the 1920s much spectroscopic evidence made it clear that the Coulomb Hamiltonian is missing certain terms. Especially for molecules containing heavier atoms, these terms, although much smaller than kinetic and Coulomb energies, are nonnegligible. These spectroscopic observations led to the introduction of a new degree of freedom for electrons and nuclei, namely spin. This empirical concept was given a theoretical basis by Paul Dirac when he introduced a relativistically correct (Lorentz covariant) form of the one-particle Schrödinger equation. The Dirac equation predicts that spin and spatial motion of a particle interact via spin-orbit coupling. In analogy spin-other-orbit coupling was introduced. The fact that particle spin has some of the characteristics of a magnetic dipole led to spin-spin coupling. Further terms without a classical counterpart are the Fermi-contact term (interaction of electronic density on a finite size nucleus with the nucleus), and nuclear quadrupole coupling (interaction of a nuclear quadrupole with the gradient of an electric field due to the electrons). Finally a parity violating term predicted by the Standard Model must be mentioned. Although it is an extremely small interaction, it has attracted a fair amount of attention in the scientific literature because it gives different energies for the enantiomers in chiral molecules.

The remaining part of this article will ignore spin terms and consider the solution of the eigenvalue (time-independent Schrödinger) equation of the Coulomb Hamiltonian.

THE SCHRÖDINGER EQUATION OF THE COULOMB HAMILTONIAN

The Coulomb Hamiltonian has a continuous spectrum due to the center of mass (COM) motion of the molecule in homogeneous space. In classical mechanics it is easy to separate off the COM motion of a system of point masses. Classically the motion of the COM is uncoupled from the other motions. The COM moves uniformly (i.e., with constant velocity) through space as if it were a point particle with mass equal to the sum M_{tot} of the masses of all the particles.

In quantum mechanics a free particle has as state function a plane wave function, which is a non-square-integrable function of well-defined momentum. The kinetic energy of this particle can take any positive value. The position of the COM is uniformly probable everywhere, in agreement with the Heisenberg uncertainty principle.

By introducing the coordinate vector **X** of the center of mass as three of the degrees of freedom of the system and eliminating the coordinate vector of one (arbitrary) particle, so that the number of degrees of freedom stays the same, one obtains by a linear transformation a new set of coordinates $\mathbf{t}_i$. These coordinates are linear combinations of the old coordinates of *all* particles (nuclei *and* electrons). By applying the chain rule one can show that

$$H = -\frac{\hbar^2}{2M_{\text{tot}}}\nabla^2_{\mathbf{X}} + H' \quad \text{with} \quad H' = -\frac{\hbar^2}{2}\sum_{i=1}^{N_{\text{tot}}-1}\frac{1}{m_i}\nabla^2_i + \frac{\hbar^2}{2M_{\text{tot}}}\sum_{i,j=1}^{N_{\text{tot}}-1}\nabla_i\cdot\nabla_j + V(\mathbf{t}).$$

The first term of H is the kinetic energy of the COM motion, which can be treated separately since H' does not depend on **X**. As just stated, its eigenstates are plane waves. The potential $V(\mathbf{t})$ consists of the Coulomb terms expressed in the new coordinates. The first term of H' has the usual appearance of a kinetic energy operator. The second term is known as the **mass polarization** term. The translationally invariant Hamiltonian H' can be shown to be self-adjoint and to be bounded from below. That is, its lowest eigenvalue is real and finite. Although H' is necessarily invariant under permutations of identical particles (since H and the COM kinetic energy are invariant), its invariance is not manifest. Not many actual molecular applications of H' exist; see, however, the seminal work on the hydrogen molecule for an early application. In the great majority of computations of molecular wavefunctions the electronic problem is solved with the clamped nucleus Hamiltonian arising in the first step of the Born–Oppenheimer approximation.

See Ref. for a thorough discussion of the mathematical properties of the Coulomb Hamiltonian. Also it is discussed in this paper whether one can arrive *a priori* at the concept of a molecule (as a stable system of electrons and nuclei with a well-defined geometry) from the properties of the Coulomb Hamiltonian alone.

CLAMPED NUCLEUS HAMILTONIAN

The clamped nucleus Hamiltonian describes the energy of the electrons in the electrostatic field of the nuclei, where the nuclei are assumed to be stationary with respect to an inertial frame. The form of the electronic Hamiltonian is

$$H = -\frac{\hbar^2}{2M_{\text{tot}}}\nabla^2_{\mathbf{X}} + H' \quad \text{with} \quad H' = -\frac{\hbar^2}{2}\sum_{i=1}^{N_{\text{tot}}-1}\frac{1}{m_i}\nabla^2_i + \frac{\hbar^2}{2M_{\text{tot}}}\sum_{i,j=1}^{N_{\text{tot}}-1}\nabla_i\cdot\nabla_j + V(\mathbf{t}).$$

The coordinates of electrons and nuclei are expressed with respect to a frame that moves with the nuclei, so that the nuclei are at rest with respect to this frame. The frame stays parallel to a space-fixed frame. It is an inertial frame because the nuclei are assumed not to be accelerated by external forces or torques. The origin of the frame is arbitrary, it is usually positioned on a central nucleus or in the nuclear center of mass. Sometimes it is stated that the nuclei are "at rest in a space-fixed frame". This statement implies that the nuclei are viewed as classical particles, because a quantum mechanical particle cannot be at rest. (It would mean that it had simultaneously zero momentum and well-defined position, which contradicts Heisenberg's uncertainty principle).

Since the nuclear positions are constants, the electronic kinetic energy operator is invariant under translation over any nuclear vector. The Coulomb potential, depending on difference vectors, is invariant as well. In the description of atomic orbitals and the computation of integrals over atomic orbitals this invariance is used by equipping all atoms in the molecule with their own localized frames parallel to the space-fixed frame.

As explained in the article on the Born–Oppenheimer approximation, a sufficient number of solutions of the Schrödinger equation of H_{el} leads to a potential energy surface (PES) $V(\mathbf{R}_1,\mathbf{R}_2,\ldots,\mathbf{R}_N)$. It is assumed that the functional dependence of V on its coordinates is such that

$$V(\mathbf{R}_1,\mathbf{R}_2,\ldots,\mathbf{R}_N) = V(\mathbf{R}'_1,\mathbf{R}'_2,\ldots,\mathbf{R}'_N)$$

for

$$\mathbf{R}'_i = \mathbf{R}_i + \mathbf{t} \ \text{(translation) and } \mathbf{R}'_i = \mathbf{R}_i + \frac{\Delta\phi}{|\mathbf{s}|}(\mathbf{s}\times\mathbf{R}_i) \ \text{(infinitesimal rotation)},$$

where **t** and **s** are arbitrary vectors and $\Delta\varphi$ is an infinitesimal angle, $\Delta\varphi >> \Delta\varphi^2$. This invariance condition on the PES is automatically fulfilled when the PES is expressed in terms of differences of, and angles between, the $\mathbf{R}_i$, which is usually the case.

Harmonic Nuclear Motion Hamiltonian

In the remaining part of this article we assume that the molecule is semi-rigid. In the second step of the BO approximation the nuclear kinetic energy T_n is reintroduced and the Schrödinger equation with Hamiltonian

$$\hat{H}_{\text{nuc}} = -\frac{\hbar^2}{2}\sum_{i=1}^{N}\sum_{\alpha=1}^{3}\frac{1}{M_i}\frac{\partial^2}{\partial R_{i\alpha}^2} + V(\mathbf{R}_1,\ldots,\mathbf{R}_N)$$

is considered. One would like to recognize in its solution: the motion of the nuclear center of mass (3 degrees of freedom), the overall rotation of the molecule (3 degrees of freedom), and the nuclear vibrations. In general, this is not possible with the given nuclear kinetic energy, because it does not separate explicitly the 6 external degrees of freedom (overall translation and rotation) from the $3N - 6$ internal degrees of freedom.

In fact, the kinetic energy operator here is defined with respect to a space-fixed (SF) frame. If we were to move the origin of the SF frame to the nuclear center of mass, then, by application of the chain rule, nuclear mass polarization terms would appear. It is customary to ignore these terms altogether and we will follow this custom.

In order to achieve a separation we must distinguish internal and external coordinates, to which end Eckart introduced conditions to be satisfied by the coordinates. We will show how these conditions arise in a natural way from a harmonic analysis in mass-weighted Cartesian coordinates.

In order to simplify the expression for the kinetic energy we introduce mass-weighted displacement coordinates

$$\boldsymbol{\rho}_i \equiv \sqrt{M_i}(\mathbf{R}_i - \mathbf{R}_i^0).$$

Since

$$\frac{\partial}{\partial \rho_{i\alpha}} = \frac{\partial}{\sqrt{M_i}(\partial R_{i\alpha} - \partial R_{i\alpha}^0)} = \frac{1}{\sqrt{M_i}}\frac{\partial}{\partial R_{i\alpha}},$$

the kinetic energy operator becomes,

$$T = -\frac{\hbar^2}{2}\sum_{i=1}^{N}\sum_{\alpha=1}^{3}\frac{\partial^2}{\partial \rho_{i\alpha}^2}.$$

If we make a Taylor expansion of V around the equilibrium geometry,

$$V = V_0 + \sum_{i=1}^{N}\sum_{\alpha=1}^{3}\left(\frac{\partial V}{\partial \rho_{i\alpha}}\right)_0 \rho_{i\alpha} + \frac{1}{2}\sum_{i,j=1}^{N}\sum_{\alpha,\beta=1}^{3}\left(\frac{\partial^2 V}{\partial \rho_{i\alpha}\partial \rho_{j\beta}}\right)_0 \rho_{i\alpha}\rho_{j\beta} + \cdots,$$

and truncate after three terms (the so-called harmonic approximation), we can describe V with only the third term. The term V_0 can be absorbed in the energy (gives a new zero of energy). The second term is vanishing because of the equilibrium condition. The remaining term contains the Hessian matrix $\mathbf{F}$ of V, which is symmetric and may be diagonalized with an orthogonal $3N \times 3N$ matrix with constant elements:

$$\mathbf{QFQ}^{\mathrm{T}} = \mathbf{\Phi} \quad \text{with} \quad \mathbf{\Phi} = \mathrm{diag}(f_1, \ldots, f_{3N-6}, 0, \ldots, 0).$$

It can be shown from the invariance of V under rotation and translation that six of the eigenvectors of $\mathbf{F}$ (last six rows of $\mathbf{Q}$) have eigenvalue zero (are zero-frequency modes). They span the *external space*. The first $3N - 6$ rows of $\mathbf{Q}$ are—for molecules in their ground state—eigenvectors with non-zero eigenvalue; they are the internal coordinates and form an orthonormal basis for a $(3N - 6)$-dimensional subspace of the nuclear configuration space $\mathbf{R}^{3N}$, the *internal space*. The zero-frequency eigenvectors are orthogonal to the eigenvectors of non-zero frequency. It can be shown that these orthogonalities are in fact the Eckart conditions. The kinetic energy expressed in the internal coordinates is the internal (vibrational) kinetic energy.

With the introduction of normal coordinates

$$q_t \equiv \sum_{i=1}^{N}\sum_{\alpha=1}^{3} Q_{t,i\alpha}\, p_{i\alpha},$$

the vibrational (internal) part of the Hamiltonian for the nuclear motion becomes in the *harmonic approximation*

$$\hat{H}_{\text{nuc}} \approx \frac{1}{2}\sum_{t=1}^{3N-6}\left[-\hbar^2\frac{\partial^2}{\partial q_t^2} + f_t q_t^2\right].$$

The corresponding Schrödinger equation is easily solved, it factorizes into $3N - 6$ equations for one-dimensional harmonic oscillators. The main effort in this approximate solution of the nuclear motion Schrödinger equation is the computation of the Hessian **F** of V and its diagonalization.

This approximation to the nuclear motion problem, described in $3N$ mass-weighted Cartesian coordinates, became standard in quantum chemistry, since the days (1980s-1990s) that algorithms for accurate computations of the Hessian **F** became available. Apart from the harmonic approximation, it has as a further deficiency that the external (rotational and translational) motions of the molecule are not accounted for. They are accounted for in a rovibrational Hamiltonian that sometimes is called *Watson's Hamiltonian.*

WATSON'S NUCLEAR MOTION HAMILTONIAN

In order to obtain a Hamiltonian for external (translation and rotation) motions coupled to the internal (vibrational) motions, it is common to return at this point to classical mechanics and to formulate the classical kinetic energy corresponding to these motions of the nuclei. Classically it is easy to separate the translational—center of mass—motion from the other motions. However, the separation of the rotational from the vibrational motion is more difficult and is not completely possible. This ro-vibrational separation was first achieved by Eckart in 1935 by imposing by what is now known as Eckart conditions. Since the problem is described in a frame (an "Eckart» frame) that rotates with the molecule, and hence is a non-inertial frame, energies associated with the fictitious forces: centrifugal and Coriolis force appear in the kinetic energy.

In general, the classical kinetic energy T defines the metric tensor $\mathbf{g} = (g_{ij})$ associated with the curvilinear coordinates $\mathbf{s} = (s_i)$ through

$$2T = \sum_{ij} g_{ij}\dot{s}_i\dot{s}_j.$$

The quantization step is the transformation of this classical kinetic energy into a quantum mechanical operator. It is common to follow Podolsky by writing down the Laplace–Beltrami operator in the same (generalized, curvilinear) coordinates **s** as used for the classical form. The equation for this operator requires the inverse of the metric tensor **g** and its determinant. Multiplication of the Laplace–Beltrami operator by $-\hbar^2$ gives the required quantum mechanical kinetic energy operator. When we apply this recipe to Cartesian coordinates, which have unit metric, the same kinetic energy is obtained as by application of the quantization rules.

The nuclear motion Hamiltonian was obtained by Wilson and Howard in 1936, who followed this procedure, and further refined by Darling and Dennison in 1940. It remained the standard until 1968, when Watson was able to simplify it drastically by commuting through the derivatives the determinant of the metric tensor. We will give the ro-vibrational Hamiltonian obtained by Watson, which often is referred to as the **Watson Hamiltonian**. Before we do this we must mention that a derivation of this Hamiltonian is also possible by starting from the Laplace operator in Cartesian form, application of coordinate transformations, and use of the chain rule. The Watson Hamiltonian, describing all motions of the N nuclei, is

$$\hat{H} = -\frac{\hbar^2}{2M_{\text{tot}}}\sum_{\alpha=1}^{3}\frac{\partial^2}{\partial X_\alpha^2} + \frac{1}{2}\sum_{\alpha,\beta=1}^{3}\mu_{\alpha\beta}(\mathcal{P}_\alpha - \Pi_\alpha)(\mathcal{P}_\beta - \Pi_\beta) + U - \frac{\hbar^2}{2}\sum_{s=1}^{3N-6}\frac{\partial^2}{\partial q_s^2} + V.$$

The first term is the center of mass term

$$\mathbf{X} \equiv \frac{1}{M_{\text{tot}}}\sum_{i=1}^{N} M_i\mathbf{R}_i \quad \text{with} \quad M_{\text{tot}} \equiv \sum_{i=1}^{N} M_i.$$

The second term is the rotational term akin to the kinetic energy of the rigid rotor. Here $\mathcal{P}_\alpha$ is the α component of the body-fixed *rigid rotor angular momentum operator*, see this article for its expression in terms of Euler angles. The operator Π_α is a component of an operator known as the *vibrational angular momentum operator* (although it does *not* satisfy angular momentum commutation relations),

$$\Pi_\alpha = -i\hbar\sum_{s,t=1}^{3N-6}\zeta_{st}^{\alpha}\, q_s \frac{\partial}{\partial q_t}$$

with the *Coriolis coupling constant*:

$$\zeta_{st}^{\alpha} = \sum_{i=1}^{N}\sum_{\beta,\gamma=1}^{3}\epsilon_{\alpha\beta\gamma}Q_{s,i\beta}\,Q_{t,i\gamma} \quad \text{and} \quad \alpha = 1,2,3.$$

Here $\varepsilon_{\alpha\beta\gamma}$ is the Levi-Civita symbol. The terms quadratic in the $\mathcal{P}_\alpha$ are centrifugal terms, those bilinear in $\mathcal{P}_\alpha$ and Π_β are Coriolis terms. The quantities $Q_{s,i\gamma}$ are the components of the normal coordinates introduced above. Alternatively, normal coordinates may be obtained by application of Wilson's GF method. The 3 × 3 symmetric matrix $\boldsymbol{\mu}$ is called the *effective reciprocal inertia tensor*. If all q_s were zero (rigid molecule) the Eckart frame would coincide with a principal axes frame (see rigid rotor) and $\boldsymbol{\mu}$ would be diagonal, with the equilibrium reciprocal moments of inertia on the diagonal. If all q_s would be zero, only the kinetic energies of translation and rigid rotation would survive.

The potential-like term U is the *Watson term*:

$$U = -\frac{1}{8}\sum_{\alpha=1}^{3}\mu_{\alpha\alpha}$$

proportional to the trace of the effective reciprocal inertia tensor.

The fourth term in the Watson Hamiltonian is the kinetic energy associated with the vibrations of the atoms (nuclei) expressed in normal coordinates q_s, which as stated above, are given in terms of nuclear displacements $\rho_{i\alpha}$ by

$$q_s = \sum_{i=1}^{N}\sum_{\alpha=1}^{3} Q_{s,i\alpha}\rho_{i\alpha} \quad \text{for} \quad s = 1,\ldots,3N-6.$$

Finally V is the unexpanded potential energy by definition depending on internal coordinates only. In the harmonic approximation it takes the form

$$V \approx \frac{1}{2} \sum_{s=1}^{3N-6} f_s q_s^2.$$

REFERENCES

1. Holgate, Sharon Ann (2009). Understanding Solid State Physics. CRC Press. pp. 177–178. ISBN 1420012320.
2. Band structure and carrier concentration
3. Electronic Band Structure (PDF). www.springer.com. Springer. p. 24. Retrieved 10 November 2016.
4. High-energy bands are important for electron diffraction physics, where the electrons can be injected into a material at high energies, see Stern, R.; Perry, J.; Boudreaux, D. (1969). "Low-Energy Electron-Diffraction Dispersion Surfaces and Band Structure in Three-Dimensional Mixed Laue and Bragg Reflections». Reviews of Modern Physics. 41 *(2): 275. Bibcode*:1969RvMP...41..275S. doi:10.1103/RevModPhys.41.275..
5. Low-energy bands are however important in the Auger effect.
6. In copper, for example, the effective mass is a tensor and also changes sign depending on the wave vector, as can be seen in the de Haas–van Alphen effect; see http://www.phys.ufl.edu/fermisurface/
7. Kittel, p. 179
8. Kittel, pp. 245-248
9. Kittel, Eq. 42 p. 267
10. Daniel Charles Mattis (1994). The Many-Body Problem: Encyclopaedia of Exactly Solved Models in One Dimension. World Scientific. p. 340. ISBN 981-02-1476-6.
11. Walter Ashley Harrison (1989). Electronic Structure and the Properties of Solids. Dover Publications. ISBN 0-486-66021-4.
12. Joginder Singh Galsin (2001). Impurity Scattering in Metal Alloys. Springer. Appendix C. ISBN 0-306-46574-4.
13. Kuon Inoue, Kazuo Ohtaka (2004). Photonic Crystals. Springer. p. 66. ISBN 3-540-20559-4.
14. Hohenberg, P; Kohn, W. (Nov 1964). "Inhomogeneous Electron Gas». Phys. Rev. 136 *(3B): B864—B871. Bibcode*:1964PhRv..136..864H. doi:10.1103/PhysRev.136. B864.
15. Paier, J.; Marsman, M.; Hummer, K.; Kresse, G.; Gerber, IC.; Angyán, JG. (Apr 2006). "Screened hybrid density functionals applied to solids.». *J* Chem Phys. 124 *(15): 154709. Bibcode*:2006JChPh.124o4709P. doi:10.1063/1.2187006. PMID 16674253

CHAPTER 10

Born–Huang Approximation

Chapter Contents

- **Introduction**
- **Mathematical Formula**
- **Consequences**
- **Upper-Bound Property**
- **Vibronic Coupling**
- **Conical Intersection**

We know that if a Hamiltonian is separable into two or more terms, then the total eigenfunctions are products of the individual eigenfunctions of the separated Hamiltonian terms, and the total eigenvalues are sums of individual eigenvalues of the separated Hamiltonian terms.

Consider, for example, a Hamiltonian which is separable into two terms, one involving coordinate q_1 and the other involving coordinate q_2.

$$\hat{H} = \hat{H}_1(q_1) + \hat{H}_2(q_2)$$

with the overall Schrödinger equation being

$$\hat{H}\psi(q_1, q_2) = E\psi(q_1, q_2)$$

INTRODUCTION

The **Born–Huang approximation** (named after Max Born and Huang Kun) is an approximation closely related to the Born–Oppenheimer approximation. It takes into account diagonal nonadiabatic effects in the electronic Hamiltonian better than the Born–Oppenheimer approximation. Despite the addition of correction terms, the electronic states remain uncoupled under the Born–Huang approximation, making it an adiabatic approximation.

MATHEMATICAL FORMULA

The Born–Huang approximation asserts that the representation matrix of nuclear kinetic energy operator in the basis of Born–Oppenheimer electronic wavefunctions is diagonal:

$$\langle \chi_{k'}(\mathbf{r};\mathbf{R})|T_{\mathrm{n}}|\chi_k(\mathbf{r};\mathbf{R})\rangle_{(\mathbf{r})} = \mathcal{T}_{\mathrm{k}}(\mathbf{R})\delta_{k'k}$$

CONSEQUENCES

The Born–Huang approximation loosens the Born–Oppenheimer approximation by including some electronic matrix elements, while at the same time maintains its diagonal structure in the nuclear equations of motion. As a result, the nuclei still move on isolated surfaces, obtained by the addition of a small correction to the Born-Oppenheimer potential energy surface.

Under the Born–Huang approximation, the Schrödinger equation of the molecular system simplifies to

$$[T_{\mathrm{n}} + E_k(\mathbf{R}) + \mathcal{T}_{\mathrm{k}}(\mathbf{R})]\ \phi_k(\mathbf{R}) = E\phi_k(\mathbf{R}) \quad \text{for} \quad k = 1,\ldots,K,$$

The quantity $[E_k(\mathbf{R}) + \mathcal{T}_{\mathrm{k}}(\mathbf{R})]$ serves as the corrected potential energy surface.

UPPER-BOUND PROPERTY

The value of Born–Huang approximation is that it provides the upper-bound for the ground state energy. The Born–Oppenheimer approximation, on the other hand, provides the lower-bound for this value.

VIBRONIC COUPLING

Vibronic coupling (also called **nonadiabatic coupling** or **derivative coupling**) in a molecule involves the interaction between electronic and nuclear vibrational motion. The term "vibronic originates from the combination of the terms vibrational and electronic . The word coupling denotes the idea that in a molecule, vibrational and electronic interactions are interrelated and influence each other, and the magnitude of vibronic coupling reflects the degree of such interrelation.

In theoretical chemistry, the vibronic coupling is neglected within the Born–Oppenheimer approximation. Vibronic couplings are crucial to the understanding of nonadiabatic processes, especially near points of conical intersections.The direct calculation of vibronic couplings is not common due to difficulties associated with its evaluation.

Vibronic coupling describes the mixing of different electronic state as a result of small vibrations.

$$\mathbf{f}_{k'k} \equiv \langle \chi_{k'}(\mathbf{r};\mathbf{R})\,|\,\hat{\nabla}_{\mathbf{R}}\chi_k(\mathbf{r};\mathbf{R})\rangle_{(\mathbf{r})}$$

Evaluation

The evaluation of vibronic coupling often involve complex mathematical treatment.

Numerical Gradients

The form of vibronic coupling is essentially derivative of the wave function. Each component of the vibronic coupling vector can be calculated with numerical differentiation methods using wave functions at displaced geometries. This is the procedure used in MOLPRO.

First order accuracy can be achieved with forward difference formula:

$$(\mathbf{f}_{k'k})_l \approx \frac{1}{d}\left[\gamma^{k'k}(\mathbf{R}|\mathbf{R}+d\mathbf{e}_l) - \gamma^{k'k}(\mathbf{R}|\mathbf{R})\right]$$

Second order accuracy can be achieved with central difference formula:

$$(\mathbf{f}_{k'k})_l \approx \frac{1}{2d}\left[\gamma^{k'k}(\mathbf{R}|\mathbf{R}+d\mathbf{e}_l) - \gamma^{k'k}(\mathbf{R}|\mathbf{R}-d\mathbf{e}_l)\right]$$

Here, $\mathbf{e}_l$ is a unit vector along direction l. $\gamma^{k'k}$ is the transition density between the two electronic states.

$$\gamma^{k'k}(\mathbf{R}_1|\mathbf{R}_2) = \langle \chi_{k'}(\mathbf{r};\mathbf{R}_1)\,|\,\chi_k(\mathbf{r};\mathbf{R}_2)\rangle_{(\mathbf{r})}$$

Evaluation of electronic wave functions for both electronic states are required at N displacement geometries for first order accuracy and 2*N displacements to achieve second order accuracy, where N is the number of nuclear degrees of freedom. This can be very expensive for larger molecules.

As with other numerical differentiation method, the evaluation of nonadiabatic coupling vector with this method is numerically unstable, limiting the accuracy of the result. Moreover, the calculation of the two transition densities in the numerator are not straightforward. The wave functions of both electronic states are expanded with Slater determinants or Configuration state functions (CSF). The contribution from the change of CSF basis is too expensive to evaluate using numerical method, and is usually ignored by employing an approximate diabatic CSF basis. This will also cause further inaccuracy of the calculated coupling vector, although the error is usually tolerable.

Analytic Gradient Methods

Evaluating derivative couplings with analytic gradient methods has the advantage of high accuracy and very low cost, usually much cheaper than one single point calculation. This means an acceleration factor of 2N. However, the process involves intense mathematical treatment and programming. As a result, few programs have currently implemented analytic evaluation of vibronic couplings. Details about this method can be found in ref. For the implementation for SA-MCSCF and MRCI in COLUMBUS please see ref.

Crossings and Avoided Crossings of Potential Energy Surfaces

Vibronic coupling is large in the case of two adiabatic potential energy surfaces coming close to each other (that is, when the energy gap between them is of the order of magnitude of one oscillation quantum). This happens in the neighbourhood of an avoided crossing of potential energy surfaces corresponding to distinct electronic states of the same spin symmetry. At the vicinity of conical intersections, where the potential energy surfaces of the same spin symmetry cross, the magnitude of vibronic coupling approaches infinity. In either case the adiabatic or Born–Oppenheimer approximation fails and vibronic couplings have to be taken into account.

The large magnitude of vibronic coupling near avoided crossings and conical intersections allows wavefunctions to propagate from one adiabatic potential energy surface to another, giving rise to nonadiabatic phenomena such as radiationless decay. The singularity of vibronic coupling at conical intersections is responsible for the existence of Berry phase, which has been discovered by Longuet-Higgins in this context.

Difficulties and Alternatives

Although crucial to the understanding of nonadiabatic processes, direct evaluation of vibronic couplings has been very limited.

Evaluation of vibronic couplings is often associated with severe difficulties in mathematical formulation and program implementations. As a result, the algorithms to evaluate vibronic couplings are not yet implemented in many quantum chemistry program suites.

The evaluation of vibronic couplings also requires correct description of at least two electronic states in regions where they are strongly coupled. This requires the use of multi-reference methods such as MCSCF and MRCI, which are very expensive and delicate methods. This is further complicated by the fact that definition of vibronic couplings requires electronic wavefunctions. Unfortunately, wavefunction based methods are usually too expensive for larger systems and popular methods for larger systems such as density functional theory and molecular mechanics cannot generate wave function information. As a result, direct evaluation of vibronic couplings are mostly limited to very small molecules. The magnitude of vibronic coupling is often introduced as an empirical parameter determined by reproducing experimental data.

Alternatively, one can avoid explicit use of derivative couplings by switch from the adiabatic to the diabatic representation of the potential energy surfaces. Although rigorous validation of a diabatic representation requires knowledge of vibronic coupling, it is often possible to construct such diabatic representations by referencing the continuity of physical quantities such as dipole moment, charge distribution or orbital occupations. However, such construction requires detailed knowledge of a molecular system and introduces significant arbitrariness. Diabatic representations constructed with different method can yield different results and the reliability of the result relies on the discretion of the researcher.

Theoretical Development

Perhaps the earliest examples of the importance of vibronic coupling were found during the 1930s. In 1934 Renner wrote about the vibronic coupling in an electronically excited Π-state in CO_2. Calculations of the lower excited levels of benzene by Sklar in 1937 (with the valence bond method) and later in 1938 by Goeppert-Mayer and Sklar (with the molecular orbital method) demonstrated a correspondence between the theoretical predictions and experimental results of the benzene spectrum. The benzene spectrum was the first qualitative computation of the efficiencies of various vibrations at inducing intensity absorption.

CONICAL INTERSECTION

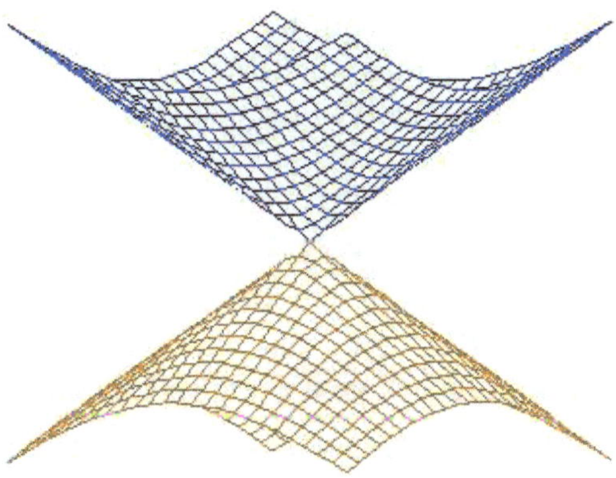

Ideal Conical Intersection.

In quantum chemistry, a **conical intersection** of two or more potential energy surfaces is the set of molecular geometry points where the potential energy surfaces are degenerate (intersect) and the non-adiabatic couplings between these states are non-vanishing. In the vicinity of conical intersections the Born–Oppenheimer approximation breaks down and the coupling between electronic and nuclear motion becomes important, allowing non-adiabatic processes to take place. The location and characterization of conical intersections are therefore essential to the understanding of a wide range of important phenomena governed by non-adiabatic events, such as photoisomerization, photosynthesis, vision and the photostability of DNA.

Conical intersections are also called **molecular funnels** or **diabolic points** as they have become an established paradigm for understanding reaction mechanisms in photochemistry as important as transitions states in thermal chemistry. This comes from the very important role they play in non-radiative de-excitation transitions from excited electronic states to the ground electronic state of molecules. For example, the stability of DNA with respect to the UV irradiation is due to such conical intersection. The molecular wave packet excited to some electronic excited state by the UVphoton follows the slope of the potential energy surface and reaches the conical intersection from above. At this point the very large vibronic coupling induces a non-radiative transition (surface-hopping) which leads the molecule back to its electronic ground state.

Degenerate points between potential energy surfaces lie in what is called the intersection or seam space with a dimensionality of 3N-8 (where N is the number of atoms). Any critical points in this space of degeneracy are characterised as minima, transition states or higher-order saddle points and can be connected to each other through the analogue of an intrinsic reaction coordinate in the seam. In benzene, for example, there is a recurrent connectivity pattern where permutationally isomeric seam segments are connected by intersections of a higher symmetry point group. The remaining two dimensions that lift the energetic degeneracy of the system are known as the branching space.

Local Characterization of Conical Intersections

Conical intersections are ubiquitous in both trivial and non-trivial chemical systems. In an ideal system of two dimensionalities, this can occur at one molecular geometry. If the potential energy surfaces are plotted as functions of the two coordinates, they form a cone centered at the degeneracy point. This is shown in the picture on the right, where the upper and lower potential energy surfaces are plotted in different colors. The name conical intersection comes from this observation.

In diatomic molecules, the number of vibrational degrees of freedom is 1. Without the necessary two dimensions required to form the cone shape, conical intersections cannot exist in these molecules. Instead, the potential energy curves experience avoided crossings.

In molecules with three or more atoms, the number of degrees of freedom for molecular vibrations is at least 3. In these systems, when spin-orbit interaction is ignored, the degeneracy of conical intersection is lifted through first order by displacements in a two dimensional subspace of the nuclear coordinate space.

The two-dimensional degeneracy lifting subspace is referred to as the **branching space** or **branching plane**. This space is spanned by two vectors, the difference of energy gradient vectors of the two intersecting electronic states (the *g vector*), and the non-adiabatic coupling vector between these two states(the *h vector*). Because the electronic states are degenerate, the wave functions of the two electronic states are subject to an arbitrary rotation. Therefore, the *g* and *h* vectors are also subject to a related arbitrary rotation, despite the fact that the space spanned by the two vectors is invariant. To enable a consistent representation of the branching space, the set of wave functions that makes the *g* and *h* vectors orthogonal is usually chosen. This choice is unique up to the signs and switchings of the two vectors, and allows these two vectors to have proper symmetry when the molecular geometry is symmetric.

The degeneracy is preserved through first order by differential displacements that are perpendicular to the branching space. The space of non-degeneracy-lifting displacements, which is the orthogonal complement of the branching space, is termed the **seam space**.

Movement within the seam space will take the molecule from one point of conical intersection to an adjacent point of conical intersection.

For an open shell molecule, when spin-orbit interaction is added to the, the dimensionality of seam space is reduced.

The presence of conical intersections is difficult to detect experimentally. Only recently it has been proposed that two-dimensional spectroscopy can be used to detect their presence through the modulation of the frequency of the vibrational coupling mode.

Categorization by Symmetry of Intersecting Electronic States

Conical intersections can occur between electronic states with the same or different point group symmetry, with the same or different spin symmetry. When restricted to a non-relativistic Coulomb Hamiltonian, conical intersections can be classified as symmetry-required, accidental symmetry-allowed, or accidental same-symmetry, according to the symmetry of the intersecting states.

A **symmetry-required** conical intersection is an intersection between two electronic states carrying the same multidimensional irreducible representation. For example, intersections between a pair of E states at a geometry that has a non-abelian group symmetry(e.g. (C3), C(3v) or D(3h)). It is named symmetry-required because these electronic states will always be degenerate as long as the symmetry is present. Symmetry-required intersections are often associated with Jahn-Teller effect.

An **accidental symmetry-allowed** conical intersection is an intersection between two electronic states that carry different point group symmetry. It is called accidental because the states may or may not be degenerate when the symmetry is present. Movement along one of the dimensions along which the degeneracy is lifted, the direction of the difference of the energy gradients of the two electronic states, will preserve the symmetry while displacements along the other degeneracy lifting dimension, the direction of the non-adiabatic couplings, will break the symmetry of the molecule. Thus, by enforcing the symmetry of the molecule, the degeneracy lifting effect caused by inter-state couplings is prevented. Therefore, the search for a symmetry-allowed intersection becomes a one-dimensional problem and does not require knowledge of the non-adiabatic couplings, significantly simplifying the effort. As a result, all the conical intersections found through quantum mechanical calculations during the early years of quantum chemistry were symmetry-allowed intersections.

An **accidental same-symmetry** conical intersection is an intersection between two electronic states that carry the same point group symmetry. While this type of intersection was traditionally more difficult to locate, a number of efficient searching algorithms and methods to compute non-adiabatic couplings have emerged in the past decade. It is now understood that same-symmetry intersections play as important a role in non-adiabatic processes as symmetry-allowed intersections.

REFERENCES

1. Todd J. Martinez (September 2010). Physical chemistry: Seaming is believing . Nature. 467: 412–413. doi:10.1038/467412a.
2. Kang, Hyuk; Kang Taek Lee; Boyong Jung; Yeon Jae Ko; Seong Keun Kim (October 2002). Intrinsic Lifetimes of the Excited State of DNA and RNA Bases . J. Am. Chem. Soc. 124 *(44): 12958–12959.* doi:10.1021/ja027627x. PMID 12405817. Retrieved 22 October 2013.
3. Lluís Blancafort (November 2010). A global picture of the S1/S0 conical intersection seam of benzene . Chemical Physics. 377: 60–65. doi:10.1016/j.chemphys.2010.08.016.
4. Matsika, Spiridoula; David R Yarkony (1 August 2001). On the effects of spin-orbit coupling on conical intersection seams in molecules with an odd number of electrons. I. Locating the seam . Journal of Chemical Physics. 115: 2038. doi:10.1063/1.1378324.
5. Farag, M. H.; T. L. C. Jansen; J. Knoester (2016). Probing the Inter-State Coupling near a Conical Intersection by Optical Spectroscopy . Journal of Physical Chemistry Letters. 7: 3328–3334. doi:10.1021/acs.jpclett.6b01463.

CHAPTER 11

Applications of Rotational Spectroscopy

Chapter Contents

- Introduction
- Applications
- Overview
- Structure of Rotational Spectra
- Quadrupole Splitting
- Stark and Zeeman Effects
- Rotational Raman Spectroscopy
- Instruments and Methods
- Rotational–Vibrational Spectroscopy

INTRODUCTION

Rotational spectroscopy is concerned with the measurement of the energies of transitions between quantized rotational states of molecules in the gas phase. The spectra of polar molecules can be measured in absorption or emission by microwave spectroscopyor by far infrared spectroscopy. The rotational spectra of non-polar molecules cannot be observed by those methods, but can be observed and measured by Raman spectroscopy. Rotational spectroscopy is sometimes referred to as *pure* rotational spectroscopy to distinguish it from rotational-vibrational spectroscopy where changes in rotational energy occur together with changes in vibrational energy, and also from ro-vibronic spectroscopy (or just vibronic spectroscopy) where rotational, vibrational and electronic energy changes occur simultaneously.

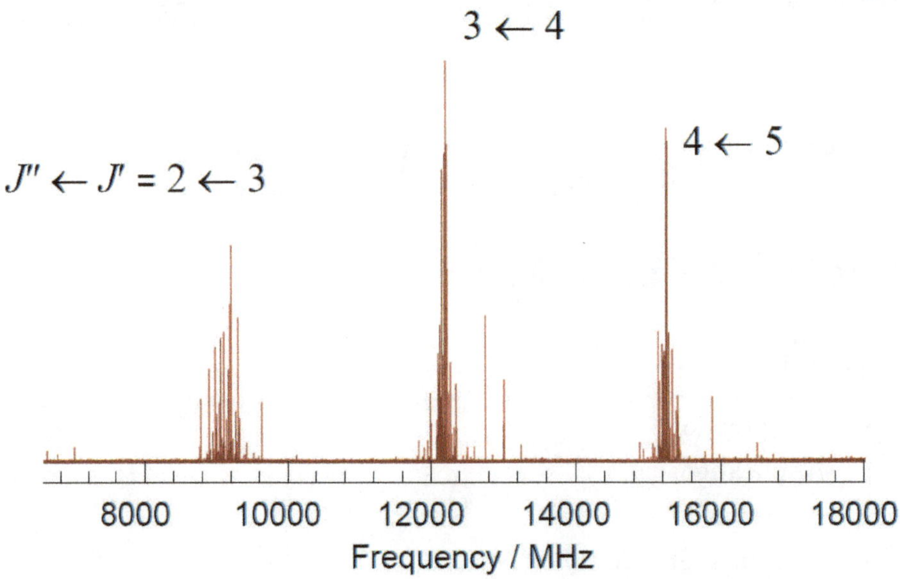

Part of the rotational spectrum of trifluoroiodomethane, CF_3I. Each rotational transition is labeled with the quantum numbers, *J*, of the final and initial states, and is extensively split by the effects of nuclear quadrupole coupling with the I nucleus.

For rotational spectroscopy, molecules are classified according to symmetry into spherical top, linear and symmetric top; analytical expressions can be derived for the rotational energy terms of these molecules. Analytical expressions can be derived for the fourth category, asymmetric top, for rotational levels up to J=3, but higher energy levels need to be determined using numerical methods. The rotational energies are derived theoretically by considering the molecules to be rigid rotors and then applying extra terms to account for centrifugal distortion, fine structure, hyperfine structure and Coriolis coupling. Fitting the spectra to the theoretical expressions gives numerical values of the angular moments of inertia from which very precise values of molecular bond lengths and angles can be derived in favorable cases. In the presence of an electrostatic field there is Stark splitting which allows molecular electric dipole moments to be determined.

An important application of rotational spectroscopy is in exploration of the chemical composition of the interstellar medium using radio telescopes.

APPLICATIONS

Rotational spectroscopy has primarily been used to investigate fundamental aspects of molecular physics. It is a uniquely precise tool for the determination of molecular structure in gas phase molecules. It can be used to establish barriers to internal rotation such as that associated with the rotation of the CH_3 group relative to the C_6H_4Cl group in chlorotoluene (C_7H_7Cl).When fine or hyperfine structure can be observed, the technique also provides information on the electronic structures of molecules. Much of current understanding of the nature of weak molecular interactions such as van der Waals, hydrogen and halogen bonds has been established through rotational spectroscopy. In connection with radio astronomy, the technique has a key role in exploration of the chemical composition of the interstellar medium. Microwave transitions are measured in the laboratory and matched to emissions from the interstellar medium using a radio telescope. NH_3 was the first stable polyatomic molecule to be identified in the interstellar medium. The measurement of chlorine monoxide is important for atmospheric chemistry. Current projects in astrochemistry involve both laboratory microwave spectroscopy and observations made using modern radiotelescopes such as the Atacama Large Millimetre Array (ALMA). Unlike NMR, Infrared and UV-Visible spectroscopies, microwave spectroscopy has not yet found widespread application in analytical chemistry.

OVERVIEW

A molecule in the gas phase is free to rotate relative to a set of mutually orthogonal axes of fixed orientation in space, centered on the center of mass of the molecule. Free rotation is not possible for molecules in liquid or solid phases due to the presence of intermolecular forces. Rotation about each unique axis is associated with a set of quantized energy levels dependent on the moment of inertia about that axis and a quantum number. Thus, for linear molecules the energy levels are described by a single moment of inertia and a single quantum number, J, which defines the magnitude of the rotational angular momentum.

For nonlinear molecules which are symmetric rotors (or symmetric tops - see next section), there are two moments of inertia and the energy also depends on a second rotational quantum number, K, which defines the vector component of rotational angular momentum along the principal symmetry axis. Analysis of spectroscopic data with the expressions detailed below results in quantitative determination of the value(s) of the moment(s) of inertia. From these precise values of the molecular structure and dimensions may be obtained.

For a linear molecule, analysis of the rotational spectrum provides values for the rotational constant and the moment of inertia of the molecule, and, knowing the atomic masses, can be used to determine the bond length directly. For diatomic molecules this process is straightforward. For linear molecules with more than two atoms it is necessary to measure the spectra of two or more isotopologues, such as $^{16}O^{12}C^{32}S$ and $^{16}O^{12}C^{34}S$. This allows a set of simultaneous equations to be set up and solved for the bond lengths). It should be noted that a bond length obtained in this way is slightly different from the equilibrium bond length. This is because there is zero-point energy in the vibrational ground state, to which the rotational states refer, whereas the equilibrium bond length is at the minimum in the potential energy curve. The relation between the rotational constants is given by

$$B_v = B - \alpha(v + \frac{1}{2})$$

where v is a vibrational quantum number and α is a vibration-rotation interaction constant which can be calculated if the B values for two different vibrational states can be found.

For other molecules, if the spectra can be resolved and individual transitions assigned both bond lengths and bond angles can be deduced. When this is not possible, as with most asymmetric tops, all that can be done is to fit the spectra to three moments of inertia calculated from an assumed molecular structure. By varying the molecular structure the fit can be improved, giving a qualitative estimate of the structure. Isotopic substitution is invaluable when using this approach to the determination of molecular structure.

Classification of Molecular Rotors

In quantum mechanics the free rotation of a molecule is quantized, so that the rotational energy and the angular momentum can take only certain fixed values, which are related simply to the moment of inertia, I, of the molecule. For any molecule, there are three moments of inertia: I_A, I_B and I_C about three mutually orthogonal axes *A*, *B*, and *C* with the origin at the center of mass of the system. The general convention, used in this article, is to define the axes such that $I_A \leq I_B \leq I_C$, corresponding to the smallest moment of inertia. Some authors, however, define the A axis as the molecular rotation axis of highest order.

The particular pattern of energy levels (and, hence, of transitions in the rotational spectrum) for a molecule is determined by its symmetry. A convenient way to look at the molecules is to divide them into four different classes, based on the symmetry of their structure. These are

- Spherical tops (spherical rotors) All three moments of inertia are equal to each other: $I_A = I_B = I_C$. Examples of spherical tops include phosphorus tetramer (P_4), carbon tetrachloride (CCl_4) and other tetrahalides, methane (CH_4), silane, (SiH_4), sulfur hexafluoride (SF_6) and other hexahalides. The molecules all belong to the cubic point groups T_d or O_h.
- Linear molecules. For a linear molecule the moments of inertia are related by $I_A << I_B = I_C$. For most purposes, I_A can be taken to be zero. Examples of linear molecules include dioxygen, O_2, dinitrogen, N_2, carbon monoxide, CO, hydroxy radical, OH, carbon dioxide, CO_2, hydrogen cyanide, HCN, carbonyl sulfide, OCS, acetylene (ethyne, HC≡CH) and dihaloethynes. These molecules belong to the point groups $C_{\infty v}$ or $D_{\infty h}$
- Symmetric tops (symmetric rotors) A symmetric top is a molecule in which two moments of inertia are the same, $I_A = I_B$ or $I_B = I_C$. By definition a symmetric top must have a 3-fold or higher order rotation axis. As a matter of convenience, spectroscopists divide molecules into two classes of symmetric tops, *Oblate symmetric tops* (saucer or disc shaped) with $I_A = I_B < I_C$ and *Prolate symmetric tops* (rugby football, or cigar shaped) with $I_A < I_B = I_C$.The spectra look rather different, and are instantly recognizable. Examples of symmetric tops include

Oblate: benzene, C_6H_6, ammonia, NH_3

Prolate: chloromethane, CH_3Cl, propyne, $CH_3C{\equiv}CH$

As a detailed example, ammonia has a moment of inertia $I_C = 4.4128 \times 10^{-47}$ kg m^2 about the 3-fold rotation axis, and moments $I_A = I_B = 2.8059 \times 10^{-47}$ kg m^2 about any axis perpendicular to the C_3 axis. Since the unique moment of inertia is larger than the other two, the molecule is an oblate symmetric top.

- Asymmetric tops (asymmetric rotors) The three moments of inertia have different values. Examples of small molecules that are asymmetric tops include water, H_2O and nitrogen dioxide, NO_2 whose symmetry axis of highest order is a 2-fold rotation axis. Most large molecules are asymmetric tops.

Selection Rules

Microwave and far-infrared Spectra

Transitions between rotational states can be observed in molecules with a permanent electric dipole moment. A consequence of this rule is that no microwave spectrum can be observed for centrosymmetric linear molecules such as N_2 (dinitrogen) or HCCH (ethyne), which are non-polar. Tetrahedral molecules such as CH_4 (methane), which have both a zero dipole moment and isotropic polarizability, would not have a pure rotation spectrum but for the effect of centrifugal distortion; when the molecule rotates about a 3-fold symmetry axis a small dipole moment is created, allowing a weak rotation spectrum to be observed by microwave spectroscopy.

With symmetric tops, the selection rule for electric-dipole-allowed pure rotation transitions is $\Delta K = 0$, $\Delta J = \pm 1$. Since these transitions are due to absorption (or emission) of a single photon with a spin of one, conservation of angular momentum implies that the molecular angular momentum can change by at most one unit. Moreover, the quantum number K is limited to have values between and including $+J$ to $-J$.

Raman Spectra

For Raman spectra the molecules undergo transitions in which an *incident* photon is absorbed and another *scattered* photon is emitted. The general selection rule for such a transition to be allowed is that the molecular polarizability must be anisotropic, which means that it is not the same in all directions. Polarizability is a 3-dimensional tensor that can be represented as an ellipsoid. The polarizability ellipsoid of spherical top molecules is in fact spherical so those molecules show no rotational Raman spectrum. For all other molecules both Stokes and anti-Stokes lines can be observed and they have similar intensities due to the fact that many rotational states are thermally populated. The selection rule for linear molecules is $\Delta J = 0, \pm 2$. The reason for the values ± 2 is that the polarizability returns to the same value twice during a rotation. The value $\Delta J = 0$ does not correspond to a molecular transition but rather to Rayleigh scattering in which the incident photon merely changes direction.

The selection rule for symmetric top molecules is

$\Delta K = 0$

If $K = 0$, then $\Delta J = \pm 2$

If $K \neq 0$, then $\Delta J = 0, \pm 1, \pm 2$

Transitions with $\Delta J = +1$ are said to belong the an R series, whereas transitions with $\Delta J = +2$ belong to an S series. Since Raman transitions involve two photons, it is possible for the molecular angular momentum to change by two units.

Units

The units used for rotational constants depend on the type of measurement. With infrared spectra in the wavenumber scale $(\tilde{\nu})$, the unit is usually the inverse centimeter, written as cm^{-1}, which is literally the number of waves in one centimeter, or the reciprocal of the wavelength in centimeters $(\tilde{\nu} = 1/\lambda)$. On the other hand, for microwave spectra in the frequency scale (ν), the unit is usually the gigahertz. The relationship between these two units is derived from the expression

$$\nu \cdot \lambda = c,$$

where ν is a frequency, λ is a wavelength and c is the velocity of light. It follows that

$$\tilde{\nu}/\text{cm}^{-1} = \frac{1}{\lambda/\text{cm}} = \frac{\nu/\text{s}^{-1}}{c/(\text{cm}\cdot\text{s}^{-1})} = \frac{\nu/\text{s}^{-1}}{2.99792458 \times 10^{10}}.$$

As 1 GHz = 10^9 Hz, the numerical conversion can be expressed as

$$\tilde{\nu}/\text{cm}^{-1} \approx \frac{\nu/\text{GHz}}{30}.$$

Effect of Vibration on Rotation

The population of vibrationally excited states follows a Boltzmann distribution, so low-frequency vibrational states are appreciably populated even at room temperatures. As the moment of inertia is higher when a vibration is excited, the rotational constants (B) decrease. Consequently, the rotation frequencies in each vibration state are different from each other. This can give rise to “satellite” lines in the rotational spectrum. An example is provided by cyanodiacetylene, H−C≡C−C≡C−C≡N.

Further, there is a fictitious force, Coriolis coupling, between the vibrational motion of the nuclei in the rotating (non-inertial) frame. However, as long as the vibrational quantum number does not change (i.e., the molecule is in only one state of vibration), the effect of vibration on rotation is not important, because the time for vibration is much shorter than the time required for rotation. The Coriolis coupling is often negligible, too, if one is interested in low vibrational and rotational quantum numbers only.

Effect of Rotation on Vibrational Spectra

Historically, the theory of rotational energy levels was developed to account for observations of vibration-rotation spectra of gases in infrared spectroscopy, which was used before microwave spectroscopy had become practical. To a first approximation, the rotation and vibration can be treated as separable, so the energy of rotation is added to the energy of vibration. For example, the rotational energy levels for linear molecules (in the rigid-rotor approximation) are

$$E_{\text{rot}} = hcBJ(J+1).$$

In this approximation, the vibration-rotation wavenumbers of transitions are

$$\tilde{\nu} = \tilde{\nu}_{\text{vib}} + B''J''(J''+1) - B'J'(J'+1),$$

where B'' and B' are rotational constants for the upper and lower vibrational state respectively, while J'' and J' are the rotational quantum numbers of the upper and lower levels. In reality, this expression has to be modified for the effects of anharmonicity of the vibrations, for centrifugal distortion and for Coriolis coupling.

For the so-called *R* branch of the spectrum, $J' = J'' + 1$ so that there is simultaneous excitation of both vibration and rotation. For the *P* branch, $J' = J'' - 1$ so that a quantum of rotational energy is lost while a quantum of vibrational energy is gained. The purely vibrational transition, $\Delta J = 0$, gives rise to the *Q* branch of the spectrum. Because of the thermal population of the rotational states the *P* branch is slightly less intense than the *R* branch.

Rotational constants obtained from infrared measurements are in good accord with those obtained by microwave spectroscopy, while the latter usually offers greater precision.

STRUCTURE OF ROTATIONAL SPECTRA

Spherical top

Spherical top molecules have no net dipole moment. A pure rotational spectrum cannot be observed by absorption or emission spectroscopy because there is no permanent dipole moment whose rotation can be accelerated by the electric field of an incident photon. Also the polarizability is isotropic, so that pure rotational transitions cannot be observed by Raman spectroscopy either. Nevertheless, rotational constants can be obtained by ro-vibrational spectroscopy. This occurs when a molecule is polar in the vibrationally excited state. For example, the molecule methane is a symmetric top but the asymmetric C-H stretching band shows rotational fine structure in the infrared spectrum, illustrated in rovibrational coupling. This spectrum is also interesting because it shows clear evidence of Coriolis coupling in the asymmetric structure of the band.

Linear Molecules

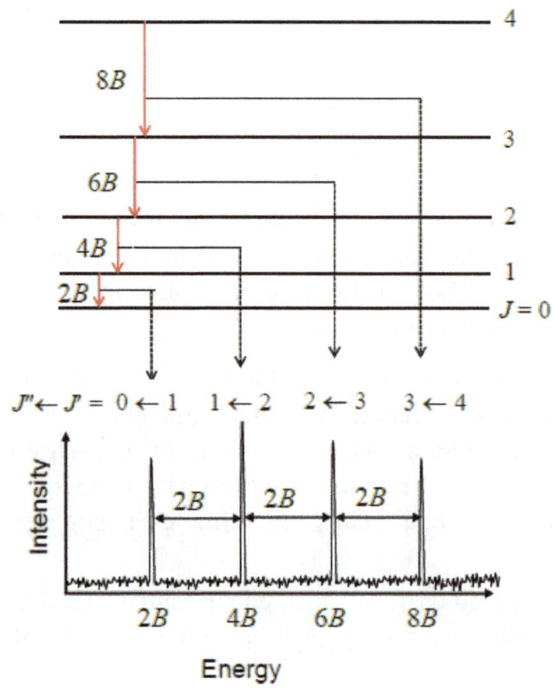

Energy levels and line positions calculated in the rigid rotor approximation.

The rigid rotor is a good starting point from which to construct a model of a rotating molecule. It is assumed that component atoms are point masses connected by rigid bonds. A linear molecule lies on a single axis and each atom moves on the surface of a sphere around the centre of mass. The two degrees of rotational freedom correspond to the spherical coordinates θ and φ which describe the direction of the molecular axis, and the quantum state is determined by two quantum numbers J and M. J defines the magnitude of the rotational angular momentum, and M its component about an axis fixed in space, such as an external electric or magnetic field. In the absence of external fields, the energy depends only on J. Under the rigid rotor model, the rotational energy levels, *F*(J), of the molecule can be expressed as,

$$F(J) = BJ(J+1) \qquad J = 0, 1, 2, \ldots$$

where B is the rotational constant of the molecule and is related to the moment of inertia of the molecule. In a linear molecule the moment of inertia about an axis perpendicular to the molecular axis is unique, that is,

$I_B = I_C, I_A = 0$, so

$$B = \frac{h}{8\pi^2 c I_B} = \frac{h}{8\pi^2 c I_C}$$

For a diatomic molecule

$$I = \frac{m_1 m_2}{m_1 + m_2} d^2$$

where m_1 and m_2 are the masses of the atoms and d is the distance between them.

Selection rules dictate that during emission or absorption the rotational quantum number has to change by unity; i.e., $\Delta J{=}1$. Thus, the locations of the lines in a rotational spectrum will be given by

$$\tilde{\nu}_{J' \leftrightarrow J''} = F(J') - F(J'') = 2B(J'' + 1) \qquad J'' = 0, 1, 2, \ldots$$

where J'' denotes the lower level and J' denotes the upper level involved in the transition.

The diagram illustrates rotational transitions that obey the $\Delta J{=}1$ selection rule. The dashed lines show how these transitions map onto features that can be observed experimentally. Adjacent $J'' \leftarrow J'$ transitions are separated by $2B$ in the observed spectrum. Frequency or wavenumber units can also be used for the x axis of this plot.

Rotational line Intensities

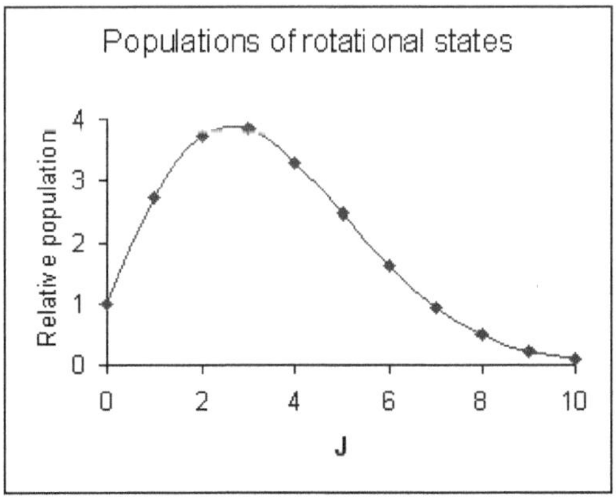

Rotational level populations with Bhc/kT = 0.05. J is the quantum number of the lower rotational state.

The probability of a transition taking place is the most important factor influencing the intensity of an observed rotational line. This probability is proportional to the population of the initial state involved in the transition. The population of a rotational state depends on two factors. The number of molecules in an excited state with quantum number J, relative to the number of molecules in the ground state, N_J/N_0 is given by the Boltzmann distribution as

$$\frac{N_J}{N_0} = e^{-\frac{E_J}{kT}} = e^{-\frac{BhcJ(J+1)}{kT}},$$

where k is the Boltzmann constant and T the absolute temperature. This factor decreases as J increases. The second factor is the degeneracy of the rotational state, which is equal to 2J+1. This factor increases as J increases. Combining the two factors

$$\text{population} \propto (2J+1)e^{-\frac{E_J}{kT}}$$

The maximum relative intensity occurs at

$$J = \sqrt{\frac{kT}{2hcB}} - 1/2$$

The diagram at the right shows an intensity pattern roughly corresponding to the spectrum above it.

Centrifugal Distortion

When a molecule rotates, the centrifugal force pulls the atoms apart. As a result, the moment of inertia of the molecule increases, thus decreasing the value of B, when it is calculated using the expression for the rigid rotor. To account for this a centrifugal distortion correction term is added to the rotational energy levels of the diatomic molecule.

$$F(J) = BJ(J+1) - DJ^2(J+1)^2 \qquad J = 0, 1, 2, \ldots$$

where D is the centrifugal distortion constant.

Therefore, the line positions for the rotational mode change to

$$\bar{\nu}_{J' \leftrightarrow J''} = 2B(J''+1) - 4D(J''+1)^3 \qquad J'' = 0, 1, 2, \ldots$$

In consequence, the spacing between lines is not constant, as in the rigid rotor approximation, but decreases with increasing rotational quantum number.

An assumption underlying these expressions is that the molecular vibration follows simple harmonic motion. In the harmonic approximation the centrifugal constant D can be derived as

$$D = \frac{h^3}{32\pi^4 I^2 r^2 kc}$$

where k is the vibrational force constant. The relationship between B and D

$$D = \frac{4B^3}{\tilde{\omega}^2}$$

where $\tilde{\omega}$ is the harmonic vibration frequency, follows. If anharmonicity is to be taken into account, terms in higher powers of J should be added to the expressions for the energy levels and line positions. A striking example concerns the rotational spectrum of hydrogen fluoride which was fitted to terms up to $[J(J+1)]^5$.

Oxygen

The electric dipole moment of the dioxygen molecule, O_2 is zero, but the molecule is paramagnetic with two unpaired electrons so that there are magnetic-dipole allowed transitions which can be observed by microwave spectroscopy. The unit electron spin has three spatial orientations with respect to the given molecular rotational angular momentum vector, K, so that each rotational level is split into three states, J = K + 1, K, and K - 1, each J state of this so-called p-type triplet arising from a different orientation of the spin with respect to the rotational motion of the molecule. The energy difference between successive J terms in any of these triplets is about 2 cm^{-1} (60 GHz), with the single exception of J = 1←0 difference which is about 4 cm^{-1}. Selection rules for magnetic dipole transitions allow transitions between successive members of the triplet (ΔJ = ±1) so that for each value of the rotational angular momentum quantum number K there are two allowed transitions. The ^{16}O nucleus has zero nuclear spin angular momentum, so that symmetry considerations demand that K have only odd values.

Symmetric Top

For symmetric rotors a quantum number J is associated with the total angular momentum of the molecule. For a given value of J, there is a $2J+1$- fold degeneracy with the quantum number, M taking the values $+J$ ...0 ... $-J$. The third quantum number, K is associated with rotation about the principal rotation axis of the molecule. In the absence of an external electrical field, the rotational energy of a symmetric top is a function of only J and K and, in the rigid rotor approximation, the energy of each rotational state is given by

$$F(J,K) = BJ(J+1) + (A-B)K^2 \qquad J = 0,1,2,\ldots \quad \text{and} \quad K = +J,\ldots 0\ldots -J$$

where $B = \dfrac{h}{8\pi^2 cI_B}$ and $A = \dfrac{h}{8\pi^2 cI_A}$ for a *prolate* symmetric top molecule or $A = \dfrac{h}{8\pi^2 cI_C}$ for an *oblate* molecule.

This gives the transition wavenumbers as

$$\tilde{\nu}_{J'\leftrightarrow J'',K} = F(J',K) - F(J'',K) = 2B(J''+1) \qquad J'' = 0,1,2,\ldots$$

which is the same as in the case of a linear molecule. With a first order correction for centrifugal distortion the transition wavenumbers become

$$\tilde{\nu}_{J'\leftrightarrow J'',K} = F(J',K) - F(J'',K) = 2\left(B - 2D_{JK}K^2\right)(J''+1) - 4D_J(J''+1)^3 \qquad J'' = 0,1,2,\ldots$$

The term in D_{JK} has the effect of removing degeneracy present in the rigid rotor approximation, with different K values.

Asymmetric Top

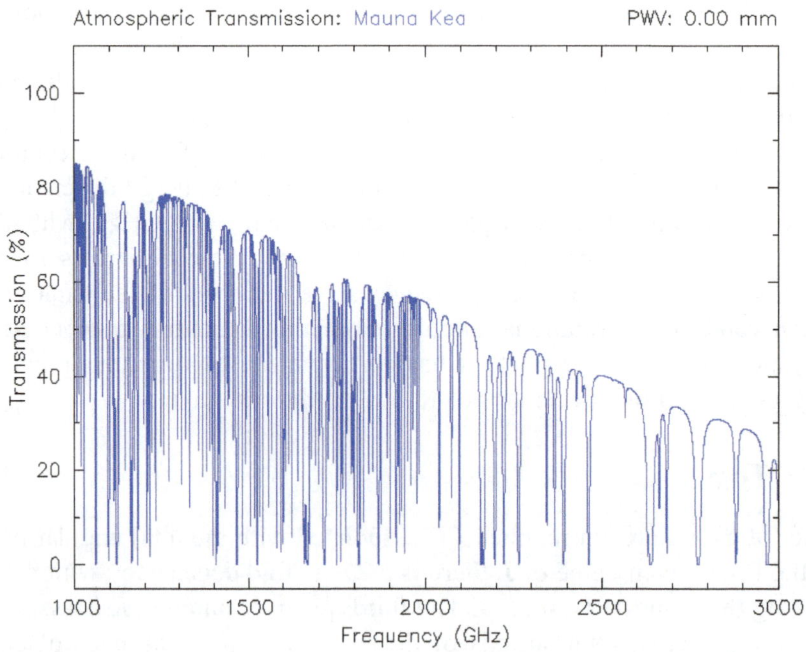

Pure rotation spectrum of atmospheric water vapour measured at Mauna Kea (33 cm^{-1} to 100 cm^{-1}).

The quantum number *J* refers to the total angular momentum, as before. Since there are three independent moments of inertia, there are two other independent quantum numbers to consider, but the term values for an asymmetric rotor cannot be derived in closed form. They are obtained by individual matrix diagonalization for each *J* value. Formulae are available for molecules whose shape approximates to that of a symmetric top.

The water molecule is an important example of an asymmetric top. It has an intense pure rotation spectrum in the far infrared region, below about 200 cm^{-1}. For this reason far infrared spectrometers have to be freed of atmospheric water vapour either by purging with a dry gas or by evacuation. The spectrum has been analyzed in detail.

QUADRUPOLE SPLITTING

When a nucleus has a spin quantum number, *I*, greater than 1/2 it has a quadrupole moment. In that case, coupling of nuclear spin angular momentum with rotational angular momentum causes splitting of the rotational energy levels. If the quantum number *J* of a rotational level is greater than *I*, $2I+1$ levels are produced; but if *J* is less than *I*, $2J+1$ levels result. The effect is one type of hyperfine splitting. For example, with ^{14}N ($I = 1$) in HCN, all levels with $J > 0$ are split into 3. The energy of the sub-levels are proportional to the nuclear quadrupole moment and a function of *F* and *J*. where *F* =

$J+I$, $J+I-1$, ..., 0, ... $|J-I|$. Thus, observation of nuclear quadrupole splitting permits the magnitude of the nuclear quadrupole moment to be determined. This is an alternative method to the use of nuclear quadrupole resonance spectroscopy. The selection rule for rotational transitions becomes

$$\Delta J = \pm 1, \Delta F = 0, \pm 1$$

STARK AND ZEEMAN EFFECTS

In the presence of a static external electric field the $2J+1$ degeneracy of each rotational state is partly removed, an instance of a Stark effect. For example, in linear molecules each energy level is split into $J+1$ components. The extent of splitting depends on the square of the electric field strength and the square of the dipole moment of the molecule. In principle this provides a means to determine the value of the molecular dipole moment with high precision. Examples include carbonyl sulfide, OCS, with $\mu = 0.71521 \pm 0.00020$ Debye. However, because the splitting depends on μ^2, the orientation of the dipole must be deduced from quantum mechanical considerations.

A similar removal of degeneracy will occur when a paramagnetic molecule is placed in a magnetic field, an instance of the Zeeman effect. Most species which can be observed in the gaseous state are diamagnetic . Exceptions are odd-electron molecules such as nitric oxide, NO, nitrogen dioxide, NO_2, some chlorine oxides and the hydroxyl radical. The Zeeman effect has been observed with dioxygen, O_2

ROTATIONAL RAMAN SPECTROSCOPY

Molecular rotational transitions can also be observed by Raman spectroscopy. Rotational transitions are Raman-allowed for any molecule with an anisotropic polarizability which includes all molecules except for spherical tops. This means that rotational transitions of molecules with no permanent dipole moment, which cannot be observed in absorption or emission, can be observed, by scattering, in Raman spectroscopy. Very high resolution Raman spectra can be obtained by adapting a Fourier Transform Infrared Spectrometer. An example is the spectrum of $^{15}N_2$. It shows the effect of nuclear spin, resulting in intensities variation of 3:1 in adjacent lines. A bond length of 109.9985 ± 0.0010 pm was deduced from the data.

INSTRUMENTS AND METHODS

The great majority of contemporary spectrometers use a mixture of commercially available and bespoke components which users integrate according to their particular needs. Instruments can be broadly categorised according to their general operating principals. Although rotational transitions can be found across a very broad region of the electromagnetic spectrum, fundamental physical constraints exist on the operational bandwidth of instrument components. It is often impractical and costly to switch to measurements within an entirely different frequency region. The instruments and operating principals described below are generally appropriate to microwave spectroscopy experiments conducted at frequencies between 6 and 24 GHz.

Absorption cells and Stark Modulation

A microwave spectrometer can be most simply constructed using a source of microwave radiation, an absorption cell into which sample gas can be introduced and a detector such as a superheterodyne receiver. A spectrum can be obtained by sweeping the frequency of the source while detecting the intensity of transmitted radiation. A simple section of waveguide can serve as an absorption cell. An important variation of the technique in which an alternating current is applied across electrodes within the absorption cell results in a modulation of the frequencies of rotational transitions. This is referred to as Stark modulation and allows the use of phase-sensitive detection methods offering improved sensitivity. Absorption spectroscopy allows the study of samples that are thermodynamically stable at room temperature. The first study of the microwave spectrum of a molecule (NH 3) was performed by Cleeton & Williams in 1934. Subsequent experiments exploited powerful sources of microwaves such as the klystron, many of which were developed for radar during the Second World War. The number of experiments in microwave spectroscopy surged immediately after the war. By 1948, Walter Gordy was able to prepare a review of the results contained in approximately 100 research papers. Commercial versions of microwave absorption spectrometer were developed by Hewlett Packard in the 1970s and were once widely used for fundamental research. Most research laboratories now exploit either Balle-Flygare or chirped-pulse Fourier transform microwave (FTMW) spectrometers.

Fourier Transform Microwave (FTMW) Spectroscopy

The theoretical framework underpinning FTMW spectroscopy is analogous to that used to describe FT-NMR spectroscopy. The behaviour of the evolving system is described by optical Bloch equations. First, a short (typically 0-3 microsecond duration) microwave pulse is introduced on resonance with a rotational transition. Those molecules that absorb the energy from this pulse are induced to rotate coherently in phase with the incident radiation. De-activation of the polarisation pulse is followed by microwave emission that accompanies decoherence of the molecular ensemble. This free induction decay occurs on a timescale of 1-100 microseconds depending on instrument settings. Following pioneering work by Dicke and co-workers in the 1950s, the first FTMW spectrometer was constructed by Ekkers and Flygare in 1975.

The Balle-Flygare FTMW Spectrometer

Balle, Campbell, Keenan and Flygare demonstrated that the FTMW technique can be applied within a "free space cell" comprising an evacuated chamber containing a Fabry-Perot cavity. This technique allows a sample to be probed only milliseconds after it undergoes rapid cooling to only a few kelvins in the throat of an expanding gas jet. This was a revolutionary development because (i) cooling molecules to low temperatures concentrates the available population in the lowest rotational energy levels. Coupled with benefits conferred by the use of a Fabry-Perot cavity, this brought a great enhancement in the sensitivity and resolution of spectrometers along with a reduction in the complexity of observed spectra; (ii) it became possible to isolate and study molecules that are very weakly bound because there is insufficient energy available for them to undergo fragmentation or chemical reaction at such low temperatures. William Klemperer was a pioneer in using this instrument for the exploration of weakly bound interactions. While the Fabry-Perot cavity of a Balle-Flygare FTMW spectrometer can typically be tuned into resonance at any frequency between 6 and 18 GHz, the bandwidth of individual measurements is restricted to about 1 MHz. An animation illustrates the operation of this instrument which is currently the most widely used tool for microwave spectroscopy.

The Chirped-Pulse FTMW Spectrometer

Noting that digitisers and related electronics technology had significantly progressed since the inception of FTMW spectroscopy, B.H. Pate at the University of Virginia designed a spectrometer which retains many advantages of the Balle-Flygare FT-MW spectrometer while innovating in (i) the use of a high speed (>4 GS/s) arbitrary waveform generator to generate a chirped microwave polarisation pulse that sweeps up to 12 GHz in frequency in less than a microsecond and (ii) the use of a high speed (>40 GS/s) oscilloscope to digitise and Fourier transform the molecular free induction decay. The result is an instrument that allows the study of weakly bound molecules but which is able to exploit a measurement bandwidth (12 GHz) that is greatly enhanced compared with the Balle-Flygare FTMW spectrometer. Modified versions of the original CP-FTMW spectrometer have been constructed by a number of groups in the United States, Canada and Europe. The instrument offers a broadband capability that is highly complementary to the high sensitivity and resolution offered by the Balle-Flygare design.

ROTATIONAL–VIBRATIONAL SPECTROSCOPY

Rotational–vibrational spectroscopy is a branch of molecular spectroscopy concerned with infrared and Raman spectra of molecules in the gas phase. Transitions involving changes in both vibrational and rotational states can be abbreviated as **rovibrational** (or **ro-vibrational**) transitions. When such transitions emit or absorb photons (electromagnetic radiation), the frequency is proportional to the difference in energy levels and can be detected by certain kinds of spectroscopy. Since changes in rotational energy levels are typically much smaller than changes in vibrational energy levels, changes in rotational state are said to give fine structure to the vibrational spectrum. For a given vibrational transition, the same theoretical treatment as for pure rotational spectroscopy gives the rotational quantum numbers, energy levels, and selection rules. In linear and spherical top molecules, rotational lines are found as simple progressions at both higher and lower frequencies relative to the pure vibration frequency. In symmetric top molecules the transitions are classified as parallel when the dipole moment change is parallel to the principal axis of rotation, and perpendicular when the change is perpendicular to that axis. The ro-vibrational spectrum of the asymmetric rotor water is important because of the presence of water vapor in the atmosphere.

Ro-vibrational spectroscopy concerns molecules in the gas phase. There are sequences of quantized rotational levels associated with both the ground and excited vibrational states. The spectra are often resolved into *lines* due to transitions from one rotational level in the ground vibrational state to one rotational level in the vibrationally excited state. The lines corresponding to a given vibrational transition form a *band.*

In the simplest cases the part of the infrared spectrum involving vibrational transitions with the same rotational quantum number ($\Delta J = 0$) in ground and excited states is called the Q-branch. On the high frequency side of the Q-branch the energy of rotational transitions is added to the energy of the vibrational transition. This is known as the R-branch of the spectrum for $\Delta J = +1$. The P-branch for $\Delta J = -1$ lies on the low wavenumber side of the Q branch. The appearance of the R-branch is very similar to the appearance of the pure rotation spectrum, and the P-branch appears as a nearly mirror image of the R-branch.

The appearance of rotational fine structure is determined by the symmetry of the molecular rotors which are classified, in the same way as for pure rotational spectroscopy, into linear molecules, spherical-, symmetric- and asymmetric- rotor classes. The quantum mechanical treatment of rotational fine structure is the same as for pure rotation.

A general convention is to label quantities that refer to the vibrational ground and excited states of a transition with double prime and single prime, respectively. For example, the rotational constant for the ground state is written as B'' and that of the excited state as B' Also, these constants are expressed in the molecular spectroscopist›s units of cm^{-1}. so that B in this article corresponds to $\bar{B} = B/hc$ in the definition of rotational constant at Rigid rotor.

Method of Combination Differences

Numerical analysis of ro-vibrational spectral data would appear to be complicated by the fact that the wavenumber for each transition depends on two rotational constants, B'' and B'. However combinations which depend on only one rotational constant are found by subtracting wavenumbers of pairs of lines (one in the P-branch and one in the R-branch) which have either the same lower level or the same upper level. For example, in a diatomic molecule the line denoted $P(J + 1)$ is due to the transition ($v = 0, J + 1$) → ($v = 1, J$), and the line $R(J - 1)$ is due to the transition ($v = 0, J - 1$) → ($v = 1, J$). The difference between the two wavenumbers corresponds to the energy difference between the ($J + 1$) and ($J - 1$) levels of the lower vibrational state and denoted by Δ_2 since it is the difference between levels differing by two units of J. If centrifugal distortion is included, it is given by

$$\Delta_2'' F(J) = \bar{\nu}[R(J-1)] - \bar{\nu}[P(J+1)] = (2B'' - 3D'')(2J+1) - D''(2J+1)^3$$

The rotational constant of the ground vibrational state B'' and centrifugal distortion constant, D'' can be found by least-squares fitting this difference as a function of J. The constant B'' is used to determine the internuclear distance in the ground state as in pure rotational spectroscopy. (See Appendix)

Similarly the difference $R(J) - P(J)$ depends only on the constants B' and D' for the excited vibrational state ($v = 1$), and B' can be used to determine the internuclear distance in that state (which is inaccessible to pure rotational spectroscopy).

$$\Delta_2' F(J) = \bar{\nu}[R(J)] - \bar{\nu}[P(J)] = (2B' - 3D')(2J+1) - D'(2J+1)^3$$

Linear Molecules

Heteronuclear Diatomic Molecules

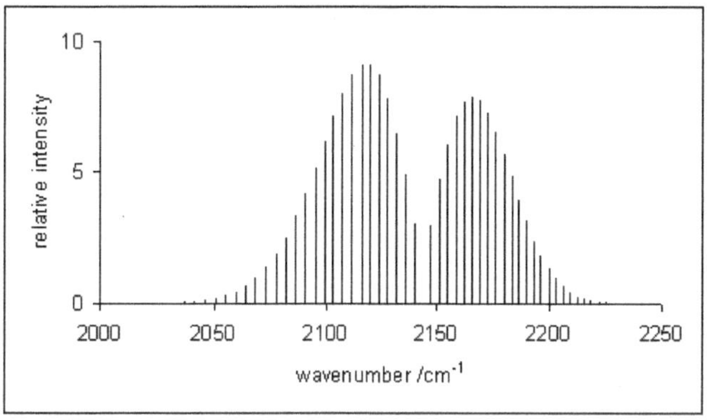

Simulated vibration-rotation line spectrum of carbon monoxide, $^{12}C^{16}O$. The P-branch is to the left of the gap near 2140 cm^{-1}, the R-branch on the right.

Diatomic molecules with the general formula AB have one normal mode of vibration involving stretching of the A-B bond. The vibrational term values $G(v)$ for an anharmonic oscillator are given, to a first approximation, by

$$G(v) = \omega_e \left(v + \frac{1}{2}\right) - \omega_e \chi_e \left(v + \frac{1}{2}\right)^2$$

where v is a vibrational quantum number, ω_e is the harmonic wavenumber and χ_e is an anharmonicity constant.

When the molecule is in the gas phase, it can rotate about an axis, perpendicular to the molecular axis, passing through the centre of mass of the molecule. The rotational energy is also quantized, with term values to a first approximation given by

$$F_v(J) = B_v J(J+1) - DJ^2(J+1)^2$$

where J is a rotational quantum number and D is a centrifugal distortion constant. The rotational constant, B_v depends on the moment of inertia of the molecule, I_v, which varies with the vibrational quantum number, v

$$B_v = \frac{h}{8\pi^2 c I_v}; \quad I_v = \frac{m_A m_B}{m_A + m_B} d_v^2$$

where m_A and m_B are the masses of the atoms A and B, and d represents the distance between the atoms. The term values of the ro-vibrational states are found (in the Born–Oppenheimer approximation) by combining the expressions for vibration and rotation.

$$G(v) + F_v(J) = \left[\omega_e\left(v+\frac{1}{2}\right) + B_v J(J+1)\right] - \left[\omega_e \chi_e \left(v+\frac{1}{2}\right)^2 + DJ^2(J+1)^2\right]$$

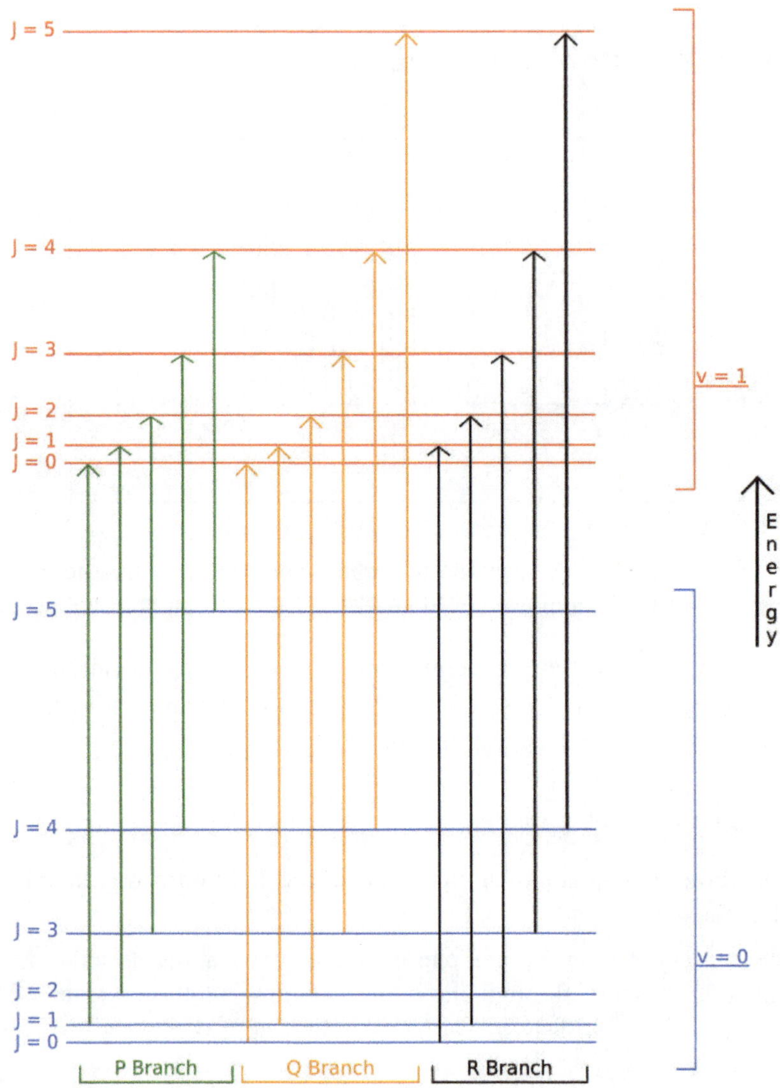

Schematic ro-vibrational energy level diagram for a linear molecule.

The first two terms in this expression correspond to a harmonic oscillator and a rigid rotor, the second pair of terms make a correction for anharmonicity and centrifugal distortion. A more general expression was given by Dunham.

The selection rule for electric dipole allowed ro-vibrational transitions, in the case of a diamagnetic diatomic molecule is $\Delta v = \pm 1\ (\pm 2, \pm 3, etc.), \Delta J = \pm 1$

The transition with Δv=±1 is known as the fundamental transition. The selection rule has two consequences.

- Both the vibrational and rotational quantum numbers must change. The transition $\Delta v = \pm 1, \Delta J = 0$ (Q-branch) is forbidden
- The energy change of rotation can be either subtracted from or added to the energy change of vibration, giving the P- and R- branches of the spectrum, respectively.

The calculation of the transition wavenumbers is more complicated than for pure rotation because the rotational constant B_v is different in the ground and excited vibrational states. A simplified expression for the wavenumbers is obtained when the centrifugal distortion constants D' and D'' are approximately equal to each other.

$$\bar{\nu} = \omega_0 + (B' + B'')m + (B' - B'')m^2 - 2(D' + D'')m^3, \quad \omega_0 = \omega_e(1 - 2\chi_e) \quad m = \pm 1, \pm 2\ etc.$$

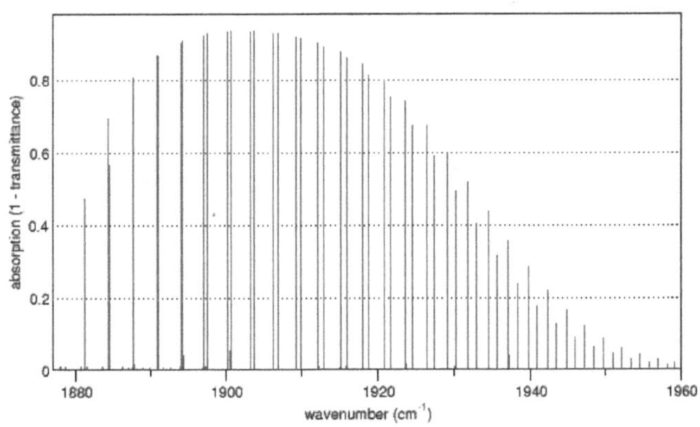

Spectrum of R-branch of nitric oxide, NO, simulated with Spectralcalc, showing λ-doubling caused by the presence of an unpaired electron in the molecule.

where positive *m* values refer to the R-branch and negative values refer to the P-branch. The term ω_0 gives the position of the (missing) Q-branch, the term $(B' + B'')m$ implies an progression of equally spaced lines in the P- and R- branches, but the third term, $(B' - B'')m^2$ shows that the separation between adjacent lines changes with changing rotational quantum number. When B'' is greater than B', as is usually the case, as *J* increases the separation between lines decreases in the R-branch and increases in the P-branch. Analysis of data from the infrared spectrum of carbon monoxide, gives value of B'' of 1.915 cm^{-1}and B'' of 1.898 cm^{-1}. The bond lengths are easily obtained from these constants as r_0 = 113.3 pm, r_1 = 113.6 pm. These bond lengths are slightly different from the equilibrium bond length. This is because there is zero-point energy in the vibrational ground state, whereas the equilibrium bond length is at the minimum in the potential energy curve. The relation between the rotational constants is given by

$$B_{\nu} = B_{eq} - \alpha\left(\nu + \frac{1}{2}\right)$$

where v is a vibrational quantum number and α is a vibration-rotation interaction constant which can be calculated when the B values for two different vibrational states can be found. For carbon monoxide r_{eq} = *113.0 pm.*

Nitric oxide, NO, is a special case as the molecule is paramagnetic, with one unpaired electron. Coupling of the electron spin angular momentum with the molecular vibration causes *lambda-doubling* with calculated harmonic frequencies of 1904.03 and 1903.68 cm^{-1}. Rotational levels are also split.

Homonuclear Diatomic Molecules

The quantum mechanics for diatomic molecules such as dinitrogen, N_2, and fluorine, F_2, is qualitatively the same as for heteronuclear diatomic molecules, but the selection rules governing transitions are different. Since the electric dipole moment of these molecules is zero, the fundamental vibrational transition is electric-dipole-forbidden. However, a weak quadrupole-allowed spectrum of N_2 can be observed when using long path-lengths both in the laboratory and in the atmosphere. The spectra of these molecules can be observed by Raman spectroscopy because the molecular vibration is Raman-allowed.

Dioxygen is a special case as the molecule is paramagnetic so magnetic-dipole-allowed transitions can be observed in the infrared. The unit electron spin has three spatial orientations with respect to the molecular rotational angular momentum vector, N, so that each rotational level is split into three states with total angular momentum (molecular rotation plus electron spin) $J\hbar$, J = N + 1, N, and N - 1, each J state of this so-called p-type triplet arising from a different orientation of the spin with respect to the rotational motion of the molecule. Selection rules for magnetic dipole transitions allow transitions between successive members of the triplet (ΔJ = ±1) so that for each value of the rotational angular momentum quantum number N there are two allowed transitions. The ^{16}O nucleus has zero nuclear spin angular momentum, so that symmetry considerations demand that N may only have odd values.

Raman Spectra of Diatomic Molecules

The selection rule is

$$\Delta J = 0, \pm 2$$

so that the spectrum has an O-branch ($\Delta J = -2$), a Q-branch ($\Delta J = 0$) and an S-branch (ΔJ=+2). In the approximation that $B'' = B' = B$ the wavenumbers are given by

$$\bar{\nu}(S(J)) = \omega_0 + 4BJ + 6B = \omega_0 + 6B, \quad \omega_0 + 10B, \quad \omega_0 + 14B, \quad \ldots$$
$$\bar{\nu}(O(J)) = \omega_0 - 4BJ + 2B = \omega_0 - 6B, \quad \omega_0 - 10B, \quad \omega_0 - 14B, \quad \ldots$$

since the S-branch starts at J=0 and the O-branch at J=2. So, to a first approximation, the separation between $S(0)$ and $O(2)$ is $12B$ and the separation between adjacent lines in both O- and S- branches is $4B$. The most obvious effect of the fact that $B'' \neq B'$ is that the Q-branch has a series of closely spaced side lines on the low-frequency side due to transitions in which $\Delta J=0$ for J=1,2 etc. Useful difference formulae, neglecting centrifugal distortion are as follows.

$$\Delta_4'' F(J) = \bar{\nu}[S(J-2)] - \bar{\nu}[O(J+2)] = 4B''(2J+1)$$
$$\Delta_4' F(J) = \bar{\nu}[S(J)] - \bar{\nu}[O(J)] = 4B'(2J+1)$$

Molecular oxygen is a special case as the molecule is paramagnetic, with two unpaired electrons.

For homonuclear diatomics, nuclear spin statistical weights lead to alternating line intensities between even$^{J''}$ and odd-$^{J''}$ levels. For nuclear spin I = 1/2 as in $^1H_2$ and $^{19}F_2$ the intensity alternation is 1:3. For $^2H_2$ and $^{14}N_2$, I=1 and the statistical weights are 6 and 3 so that the even-$^{J''}$ levels are twice as intense. For $^{16}O_2$ (I=0) all transitions with even values of $^{J''}$ are forbidden.

Polyatomic Linear Molecules

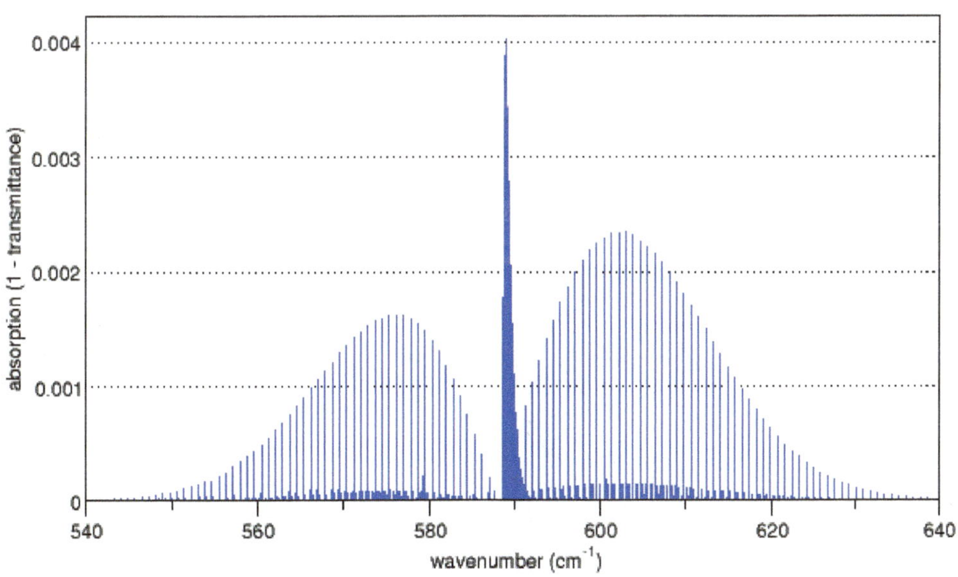

Spectrum of bending mode in $^{14}N^{14}N^{16}O$ simulated with Spectralcalc. The weak superimposed spectrum is due to species containing ^{15}N at natural abundance of 0.3%

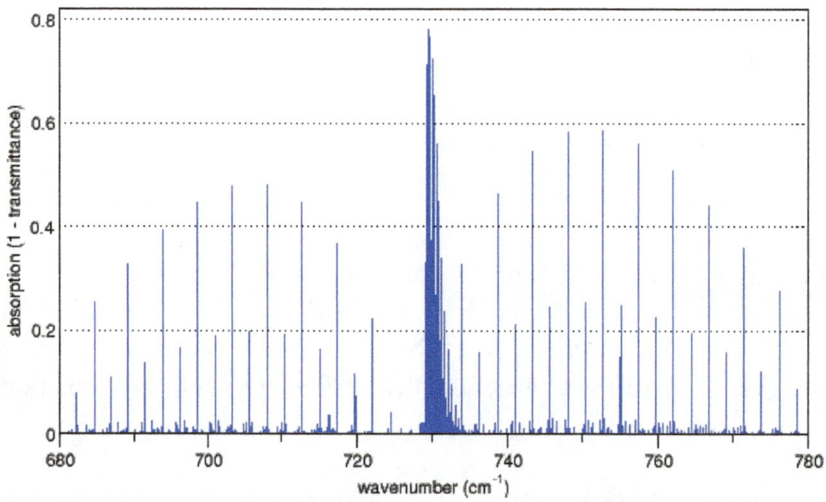

Spectrum of a perpendicular band from acetylene, C_2H_2, simulated with Spectralcalcshowing 1,3 intensity alternation in both P- and R- branches. See also Hollas p157

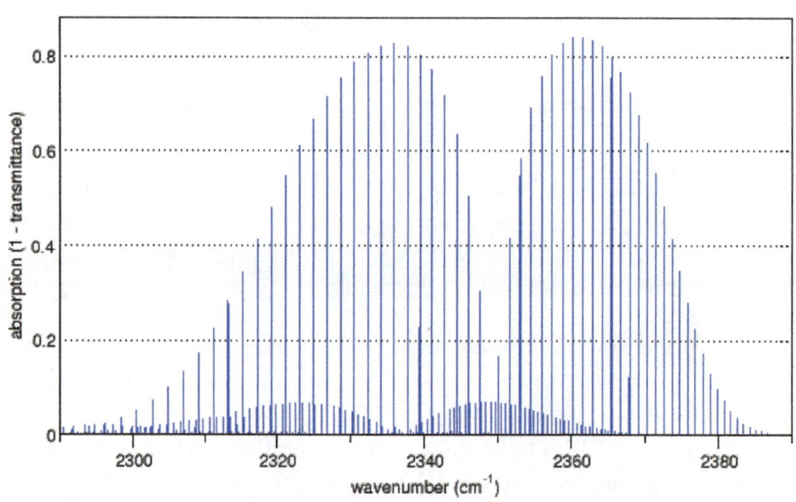

Spectrum of the asymmetric stretching (parallel) band of carbon dioxide, $^{12}C^{16}O_2$simulated with Spectralcalc. The weak superimposed spectrum is due to absorption of the first vibrationally excited level (0 1^1 0), which due to its low energy is populated at room temperature

These molecules fall into two classes, according to symmetry: centrosymmetric molecules with point group $D_{\infty h}$, such as carbon dioxide, CO_2, and ethyne or acetylene, HCCH; and non-centrosymmetric molecules with point group $C_{\infty v}$ such as hydrogen cyanide, HCN, and nitrous oxide, NNO. Centrosymmetric linear molecules have a dipole moment of zero, so do not show a pure rotation spectrum in the infrared or microwave regions. On the other hand, in certain vibrational excited states the molecules do have a dipole moment so that a ro-vibrational spectrum can be observed in the infrared.

The spectra of these molecules are classified according to the direction of the dipole moment change vector. When the vibration induces a dipole moment change pointing along the molecular axis the term *parallel* is applied, with the symbol $\parallel$. When the vibration induces a dipole moment pointing perpendicular to the molecular axis the term *perpendicular* is applied, with the symbol $\perp$. In both cases the P- and R- branch wavenumbers follow the same trend as in diatomic molecules. The two classes differ in the selection rules that apply to ro-vibrational transitions. For parallel transitions the selection rule is the same as for diatomic molecules, namely, the transition corresponding to the Q-branch is forbidden. An example is the C-H stretching mode of hydrogen cyanide.

For a perpendicular vibration the transition $\Delta J=0$ is allowed. This means that the transition is allowed for the molecule with the same rotational quantum number in the ground and excited vibrational state, for all the populated rotational states. This makes for an intense, relatively broad, Q-branch consisting of overlapping lines due to each rotational state. The N-N-O bending mode of nitrous oxide, at ca. 590 cm^{-1} is an example.

The spectra of centrosymmetric molecules exhibit alternating line intensities due to quantum state symmetry effects, since rotation of the molecule by 180° about a 2-fold rotation axis is equivalent to exchanging identical nuclei. In carbon dioxide, the oxygen atoms of the predominant isotopic species $^{12}C^{16}O_2$ have spin zero and are bosons, so that the total wavefunction must be symmetric when the two ^{16}O nuclei are exchanged. The nuclear spin factor is always symmetric for two spin-zero nuclei, so that the rotational factor must also be symmetric which is true only for even-J levels. The odd-J rotational levels cannot exist and the allowed vibrational bands consist of only absorption lines from even-J initial levels. The separation between adjacent lines in the P- and R- branches is close to 4B rather than 2B as alternate lines are missing. For acetylene the hydrogens of $^1H^{12}C^{12}C^1H$ have spin ½ and are fermions, so the total wavefunction is antisymmetric when two ^1H nuclei are exchanged. As is true for ortho and para hydrogen the nuclear spin function of the two hydrogens has three symmetric ortho states and one antisymmetric para states. For the three ortho states, the rotational wave function must be antisymmetric corresponding to odd J, and for the one para state it is symmetric corresponding to even J. The population of the odd J levels are therefore three times higher than the even J levels, and alternate line intensities are in the ratio 3:1.

Spherical Top Molecules

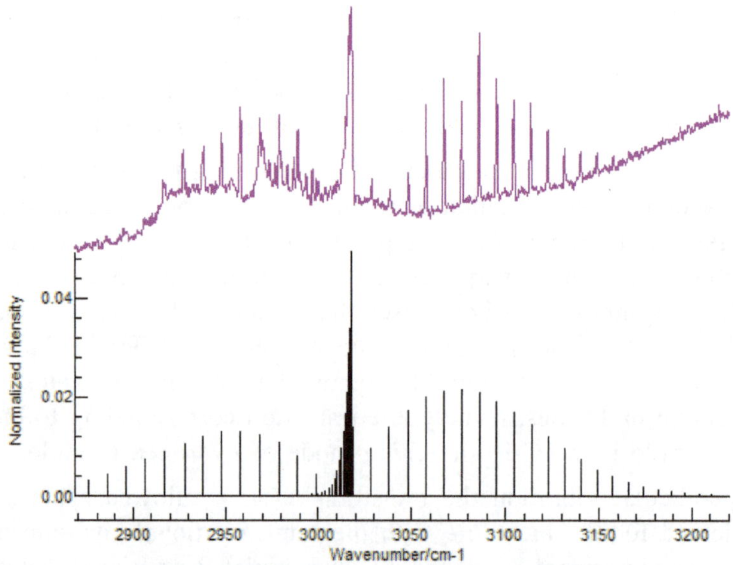

Upper: low-resolution infrared spectrum of asymmetric stretching band of methane (CH_4); lower: PGOPHER simulated line spectrum, ignoring Coriolis coupling.

These molecules have equal moments of inertia about any axis, and belong to the point groups T_d (tetrahedral AX_4) and O_h (octahedral AX_6). Molecules with these symmetries have a dipole moment of zero, so do not have a pure rotation spectrum in the infrared or microwave regions.

Tetrahedral molecules such as methane, CH_4, have infrared-active stretching and bending vibrations, belonging to the T_2 (sometimes written as F_2) representation. These vibrations are triply degenerate and the rotational energy levels have three components separated by the Coriolis interaction. The rotational term values are given, to a first order approximation, by

$$F^+ = B_\nu J(J+1) + 2B_\nu\zeta_r(J+1)$$
$$F^0 = B_\nu J(J+1)$$
$$F^- = B_\nu J(J+1) - 2B_\nu\zeta_r(J+1)$$

where ζ_r is a constant for Coriolis coupling. The selection rule for a fundamental vibration is

$$\Delta J = 0, \pm 1$$

Thus, the spectrum is very much like the spectrum from a perpendicular vibration of a linear molecule, with a strong Q-branch composed of many transitions in which the rotational quantum number is the same in the vibrational ground and excited states, $J' = J'' = 1, 2...$ The effect of Coriolis coupling is clearly visible in the C-H stretching vibration of methane, though detailed study has shown that the first-order formula for Coriolis coupling, given above, is not adequate for methane.

Symmetric Top Molecules

These molecules have a unique principal rotation axis of order 3 or higher. There are two distinct moments of inertia and therefore two rotational constants. For rotation about any axis perpendicular to the unique axis, the moment of inertia is $I_\perp$ and the rotational constant is $B = \frac{h}{8\pi^2 c I_\perp}$, as for linear molecules. For rotation about the unique axis, however, the moment of inertia is $I_\parallel$ and the rotational constant is $A = \frac{h}{8\pi^2 c I_\parallel}$. Examples include ammonia, NH_3 and methyl chloride, CH_3Cl (both of molecular symmetry described by point group C_{3v}), boron trifluoride, BF_3 and phosphorus pentachloride, PCl_5 (both of point group D_{3h}), and benzene, C_6H_6 (point group D_{6h}).

For symmetric rotors a quantum number *J* is associated with the total angular momentum of the molecule. For a given value of J, there is a 2*J*+1- fold degeneracy with the quantum number, *M* taking the values +*J* ...0 ... -*J*. The third quantum number, *K* is associated with rotation about the principal rotation axis of the molecule. As with linear molecules, transitions are classified as *parallel*, $\parallel$ or perpendicular, $\perp$, in this case according to the direction of the dipole moment change with respect to the principal rotation axis. A third category involves certain overtones and combination bands which share the properties of both parallel and perpendicular transitions. The selection rules are

$\parallel$ If $K \neq 0$, then $\Delta J = 0, \pm 1$ and $\Delta K = 0$

If $K = 0$, then $\Delta J = \pm 1$ and $\Delta K = 0$

$\perp$ $\Delta J = 0, \pm 1$ and $\Delta K = \pm 1$

The fact that the selection rules are different is the justification for the classification and it means that the spectra have a different appearance which can often be immediately recognized. An expression for the calculated wavenumbers of the P- and R- branches may be given as

$$\begin{aligned}\bar{\nu} &= \bar{\nu}_0 + (B' + B'')m + (B' - B'' - D'_J + D''_J)m^2 \\ &-2(D'_J + D''_J)m^3 - (D'_J - D''_J)m^4 \\ &+ \{[(A' - B') - (A'' - B'')] - [D'_{JK} + D''_{JK}]\,m - [D'_{JK} - D''_{JK}]\,m^2\}\,K^2 - (D'_K - D''_K)K^4\end{aligned}$$

in which $m = J+1$ for the R-branch and -*J* for the P-branch. The three centrifugal distortion constants D_J, D_{JK}, and D_K are needed to fit the term values of each level. The wavenumbers of the sub-structure corresponding to each band are given by

$$\bar{\nu} = \bar{\nu}_{sub} + (B' - B'')J(J+1) - (D'_J - D''_J)J^2(J+1)^2 - (D'_{JK} - D''_{JK})J(J+1)K^2$$

$\bar{\nu}_{sub}$ represents the Q-branch of the sub-structure, whose position is given by

$$\bar{\nu}_{sub} = \bar{\nu}_0 + [(A' - B') - (A'' - B'')]\,K^2 - (D'_K - D''_K)K^4.$$

Parallel Bands

The C-Cl stretching vibration of methyl chloride, CH_3Cl, gives a parallel band since the dipole moment change is aligned with the 3-fold rotation axis. The line spectrum shows the sub-structure of this band rather clearly; in reality, very high resolution spectroscopy would be needed to resolve the fine structure fully. Allen and Cross show parts of the spectrum of CH_3D and give a detailed description of the numerical analysis of the experimental data.

Perpendicular Bands

The selection rule for perpendicular bands give rise to more transitions than with parallel bands. A band can be viewed as a series of sub-structures, each with P, Q and R branches. The Q-branches are separated by approximately $2(A'-B')$. The asymmetric HCH bending vibration of methyl chloride is typical. It shows a series of intense Q-branches with weak rotational fine structure. Analysis of the spectra is made more complicated by the fact that the ground-state vibration is bound, by symmetry, to be a degenerate vibration, which means that Coriolis coupling also affects the spectrum.

Hybrid Bands

Overtones of a degenerate fundamental vibration have components of more than one symmetry type. For example, the first overtone of a vibration belonging to the E representation in a molecule like ammonia, NH_3, will have components belonging to A_1 and E representations. A transition to the A_1 component will give a parallel band and a transition to the E component will give perpendicular bands; the result is a hybrid band.

Inversion in Ammonia

For ammonia, NH_3, the symmetric bending vibration is observed as two branches near 930 cm^{-1} and 965 cm^{-1}. This so-called inversion doubling arises because the symmetric bending vibration is actually a large-amplitude motion known as inversion, in which the nitrogen atom passes through the plane of the three hydrogen atoms, similar to the inversion of an umbrella. The potential energy curve for such a vibration has a double minimum for the two pyramidal geometries, so that the vibrational energy levels occur in pairs which correspond to combinations of the vibrational states in the two potential minima. The two $v = 1$ states combine to form a symmetric state (1^+) at 932.5 cm^{-1} above the ground (0^+) state and an antisymmetric state (1^-) at 968.3 cm^{-1}.

The vibrational ground state ($v = 0$) is also doubled although the energy difference is much smaller, and the transition between the two levels can be measured directly in the microwave region, at ca. 24 Ghz (0.8 cm^{-1}). This transition is historically significant and was used in the ammonia MASER, the fore-runner of the LASER.

Asymmetric Top Molecules

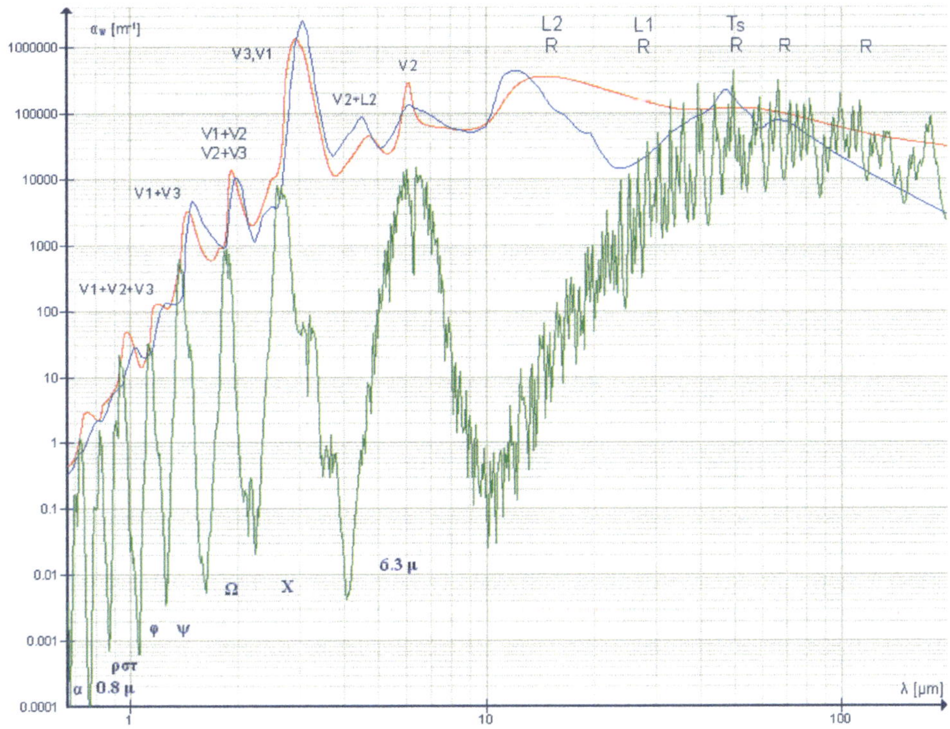

Absorption spectrum (attenuation coefficient vs. wavelength) of liquid water (red) atmospheric water vapor (green) and ice (blue line) between 667 nm and 200 μm. The plot for vapor is a transformation of data *Synthetic spectrum for gas mixture "Pure H_2O"* (296K, 1 atm) retrieved from Hitran on the Web Information System.

Asymmetric top molecules have at most one or more 2-fold rotation axes. There are three unequal moments of inertia about three mutually perpendicular principal axes. The spectra are very complex. The transition wavenumbers cannot be expressed in terms of an analytical formula but can be calculated using numerical methods.

The water molecule is an important example of this class of molecule, particularly because of the presence of water vapor in the atmosphere. The low-resolution spectrum shown in green illustrates the complexity of the spectrum. At wavelengths greater than 10 μm (or wavenumbers less than 1000 cm^{-1}) the absorption is due to pure rotation. The band around 6.3 μm (1590 cm^{-1}) is due to the HOH bending vibration; the considerable breadth of this band is due to the presence of extensive rotational fine structure. High-resolution spectra of this band are shown in Allen and Cross, p 221. The symmetric and asymmetric stretching vibrations are close to each other, so the rotational fine structures of these bands overlap. The bands at shorter wavelength are overtones and combination bands, all of which show rotational fine structure. Medium resolution spectra of the bands around 1600 cm^{-1} and 3700 cm^{-1} are shown in Banwell and McCash, p91.

Ro-vibrational bands of asymmetric top molecules are classed as A-, B- or C- type for transitions in which the dipole moment change is along the axis of smallest moment of inertia to the highest.

Experimental Methods

Ro-vibrational spectra are usually measured at high spectral resolution. In the past, this was achieved by using an echelle grating as the spectral dispersion element in a grating spectrometer. This is a type of diffraction grating optimized to use higher diffraction orders. The resolving power of an FTIR spectrometer depends on the maximum retardation of the moving mirror. For example, to achieve a resolution of 0.1 cm^{-1}, the moving mirror must have a maximum displacement of 10 cm from its position at zero path difference. Connes measured the vibration-rotation spectrum of Venusian CO_2 at this resolution. A spectrometer with 0.001 cm^{-1} resolution is now available commercially. The throughput advantage of FTIR is important for high-resolution spectroscopy as the monochromator in a dispersive instrument with the same resolution would have very narrow entrance and exit slits.

When measuring the spectra of gases it is relatively easy to obtain very long path-lengths by using a multiple reflection cell.This is important because it allows the pressure to be reduced so as to minimize pressure broadening of the spectral lines, which may degrade resolution. Path lengths up to 20m are commercially available.

REFERENCES

1. Gordy, W. (1970). A. Weissberger, ed. Microwave Molecular Spectra in Technique of Organic Chemistry. **IX**. New York: Interscience.
2. Nair, K.P.R.; Demaison, J.; Wlodarczak, G.; Merke, I. (236). Millimeterwave rotational spectrum and internal rotation in o-chlorotoluene. Journal of Molecular Spectroscopy. **237** *(2): 137–142. Bibcode*:2006JMoSp.237..137N. doi:10.1016/j.jms.2006.03.011.
3. Cheung, A.C.; Rank, D.M.; Townes, C.H.; Thornton, D.D. & Welch, W.J. (1968). Detection of NH3 molecules in the interstellar medium by their microwave emission spectra. Physical Review Letters. **21** *(25): 1701–5. Bibcode*:1968PhRvL..21.1701C. doi:10.1103/PhysRevLett.21.1701.
4. Ricaud, P.; Baron, P; de La Noë, J. (2004). Quality assessment of ground-based microwave measurements of chlorine monoxide, ozone, and nitrogen dioxide from the NDSC radiometer at the Plateau de Bure. Ann. Geophys. **22**: 1903–15. Bibcode:2004AnGeo..22.1903R. doi:10.5194/angeo-22-1903-2004.
5. Astrochemistry in Virginia. Retrieved 2 December 2012.
6. Atkins & de Paula 2006, p. 444
7. Banwell & McCash 1994, p. 99
8. Moment of inertia values from Atkins & de Paula 2006, p. 445
9. Hollas 1996, p. 95

10. Hollas 1996, p. 104 shows part of the observed rotational spectrum of silane
11. Atkins & de Paula 2006, p. 447
12. Banwell & McCash 1994, p. 49
13. Hollas 1996, p. 111
14. Atkins & de Paula 2006, pp. 474–5
15. Banwell & McCash 1994, Section 4.2, p. 105, Pure Rotational Raman Spectra
16. Alexander, A. J.; Kroto, H. W.; Walton, D. R. M. (1967). The microwave spectrum, substitution structure and dipole moment of cyanobutadiyne. J. Mol. Spectrosc. **62**: 175–180. Bibcode:1976JMoSp..62..175A. doi:10.1016/0022-2852(76)90347-7. Illustrated in Hollas 1996, p. 97
17. Banwell & McCash 1994, p. 63.
18. Banwell & McCash 1994, p. 40
19. Atkins & de Paula 2006, p. 449
20. Banwell & McCash 1994, p. 45
21. Jennings, D.A.; Evenson, K.M; Zink, L.R.; Demuynck, C.; Destombes, J.L.; Lemoine, B; Johns,J.W.C. (April 1987). High-resolution spectroscopy of HF from 40 to 1100 cm^{-1}: Highly accurate rotational constants". Journal of Molecular Spectroscopy. **122** *(2): 477–480. Bibcode*:1987JMoSp.122..477J. doi:10.1016/0022-2852(87)90021-X.pdf
22. Strandberg, M. W. P.; Meng, C. Y.; Ingersoll, J. G. (1949). The Microwave Absorption Spectrum of Oxygen. Phys.Rev. **75** *(10): 1524–8. Bibcode*:1949PhRv...75.1524S. doi:10.1103/PhysRev.75.1524.pdf
23. Krupenie, Paul H. (1972). The Spectrum of Molecular Oxygen (PDF). J. Phys. Chem. Ref. Data 1, 423 (1972). **1** *(2): 423–534. Bibcode*:1972JPCRD...1..423K. doi:10.1063/1.3253101.
24. Hollas 1996, p. 101
25. Hollas 1996, p. 102 shows the effect on the microwave spectrum of H 3SiNCS.
26. Hollas 1996, p. 103
27. Hall, Richard T.; Dowling, Jerome M. (1967). Pure Rotational Spectrum of Water Vapor. J. Chem. Phys. **47** *(7): 2454–61. Bibcode*:1967JChPh..47.2454H. doi:10.1063/1.1703330. Hall, Richard T.; Dowling, Jerome M. (1971). Erratum: Pure Rotational Spectrum of Water Vapor. J. Chem. Phys. **54** *(11): 4968. Bibcode*:1971JChPh..54.4968H. doi:10.1063/1.1674785.
28. Simmons, James W.; Anderson, Wallace E.; Gordy,Walter (1950). Microwave Spectrum and Molecular Constants of Hydrogen Cyanide. Phys. Rev. **77**: 77–79. Bibcode:1950PhRv...77...77S. doi:10.1103/PhysRev.77.77.
29. Chang, Raymond (1971). Basic Principles of Spectroscopy. McGraw-Hill. p139
30. Hollas 1996, p. 102 gives the equations for diatomic molecules and symmetric tops

31. Hollas 1996, p. 102

32. Burkhalter, James H.; Roy S. Anderson; William V. Smith; Walter Gordy (1950). The Fine Structure of the Microwave Absorption Spectrum of Oxygen. Phys. Rev. **79** *(4): 651–5. Bibcode*:1950PhRv...79..651B. doi:10.1103/PhysRev.79.651.

33. Hollas 1996, p. 113, illustrates the spectrum of 15N 2 obtained using 476.5 nm radiation from an argon ion laser.

34. Cleeton, C.E.; Williams, N.H. (1934). Electromagnetic waves of 1.1 cm wavelength and the absorption spectrum of ammonia. Physical Review. **45** *(4): 234–7. Bibcode*:1934PhRv...45..234C. doi:10.1103/PhysRev.45.234.

35. Gordy, W. (1948). Microwave spectroscopy. Reviews of Modern Physics. **20** *(4): 668–717. Bibcode*:1948RvMP...20..668G. doi:10.1103/RevModPhys.20.668.

36. June 1971, Hewlett Packard Journal (PDF). Retrieved November 2012. Check date values in: |access-date= (help)

37. Schwendemann, R.H. (1978). Transient Effects in Microwave Spectroscopy. Annu. Rev. Phys. Chem. **29**: 537–558. Bibcode:1978ARPC...29..537S. doi:10.1146/annurev.pc.29.100178.002541.

38. Dicke, R.H.; Romer, R.H. (1955). Pulse Techniques in Microwave Spectroscopy. Rev. Sci. Inst. **26** *(10): 915–928. Bibcode*:1955RScI...26..915D. doi:10.1063/1.1715156.

39. Ekkers, J.; Flygare, W.H. (1976). Pulsed microwave Fourier transform spectrometer. Rev. Sci. Inst. **47** *(4): 448–454. Bibcode*:1976RScI...47..448E. doi:10.1063/1.1134647.

40. Balle, T.J.; Campbell, E.J.; Keenan, M.R.; Flygare, W.H. (1980). A new method for observing the rotational spectra of weak molecular complexes: KrHCl. J. Chem. Phys. **72** *(2): 922–932. Bibcode*:1980JChPh..72..922B. doi:10.1063/1.439210.

41. Jager, W. Balle-Flygare FTMW spectrometer animation.

42. Web page of B.H. Pate Research Group, Department of Chemistry, University of Virginia. Retrieved November 2012. Check date values in: |access-date= (help)

43. Brown, G.G.; Dian, B.C.; Douglass, K.O.; Geyer, S.M.; Pate, B.H. (2006). The rotational spectrum of epifluorohydrin measured by chirped-pulse Fourier transform microwave spectroscopy. J. Mol. Spectrosc. **238**: 200–212. Bibcode:2006JMoSp.238..200B. doi:10.1016/j.jms.2006.05.003.

44. Grubbs, G.S.; Dewberry, C.T.; Etchison, K.C.; Kerr, K.E.; Cooke, S.A. (2007). A search accelerated correct intensity Fourier transform microwave spectrometer with pulsed laser ablation source. Rev. Sci. Inst. **78** *(9): 096106. Bibcode*:2007RScI...78i6106G. doi:10.1063/1.2786022.

45. Wilcox, D.S.; Hotopp, K.M.; Dian, B.C. (2011). Two-Dimensional Chirped-Pulse Fourier Transform Microwave Spectroscopy. J. Phys. Chem. A. **115** *(32): 8895–8905.* doi:10.1021/jp2043202.

Index

A

Antibonding electrons 93, 99, 106
Atomic orbital 55, 56, 58, 59, 63, 74, 75
Atomic orbital (AO) 91
Atomic orbitals (AOs) 50

D

Density functional theory (DFT) 82
Dissociation energy 96

E

Electron energies 60

F

Formation of bonds 38
Fundamental frequency 57

H

Heisenberg's uncertainty principle 28, 31
Hückel molecular orbital (HMO) 83
Hydrogen fluoride (HF) 100

I

Idealized wavefunctions 31
Independent-particle model 57
Inherent ambiguity 31

L

Linear combination of atomic orbitals (LCAO) 83, 91
Linear combinations of atomic orbitals (LCAO) 82, 87

M

Many-worlds interpretation (MWI) 30
Molecular orbital (MO) 38, 82, 86, 91
Molecular orbitals (MOs) 50
Molecular orbital theory 58
Molecule 38, 39, 40, 41, 42, 44, 45, 46, 47, 49, 50

N

Natural Bond Orbitals (NBO) 51
Neutron diffraction 31

O

Orbital energy 91, 101, 103
Orbital slightly 96

P

Pauli Exclusion Principle 58
Photoelectron spectrum 94, 97, 103
Plum pudding model 60

Potential energy surface (PES) 212

Q

Quantum field theory 18, 28
Quantum theory of atoms in molecules (QTAIM) 75

S

Saturnian Model 60
Self-consistent field (SCF) 86
Sigma orbital 97
Singlet oxygen 98
Somewhat arbitrarily 56
Spectral lines 57, 60

V

Valence bond 38, 39, 40, 41, 42, 49, 50
Valence bond (VB) 84